FOOD AND BEVERAGE CONSUMPTION AND HEALTH

FINGER MILLET

A VALUED CEREAL

FOOD AND BEVERAGE CONSUMPTION AND HEALTH

Additional books in this series can be found on Nova's website under the Series tab.

Additional e-books in this series can be found on Nova's website under the e-book tab.

NUTRITION AND DIET RESEARCH PROGRESS

Additional books in this series can be found on Nova's website under the Series tab.

Additional e-books in this series can be found on Nova's website under the e-book tab.

FOOD AND BEVERAGE CONSUMPTION AND HEALTH

FINGER MILLET

A VALUED CEREAL

K. S. PREMAVALLI
EDITOR

Nova Science Publishers, Inc.
New York

For permission to use material from this book please contact us:
Telephone 631-231-7269; Fax 631-231-8175
Web Site: http://www.novapublishers.com

NOTICE TO THE READER

LIBRARY OF CONGRESS CATALOGING-IN-PUBLICATION DATA

Library of Congress Control Number: 2012933170

ISBN: 978-1-62081-224-2

Published by Nova Science Publishers, Inc. † New York

This book is dedicated to
Defence Research and Development Organisation
Government of India
Ministry of Defence, Delhi, India

CONTENTS

PREFACE

Finger millet is most promising among the millets for its strength of minerals, specifically calcium, iron and dietary fibre, polyphenols. These compounds prove to be good for skeletal health and possess antidiabetic properties. Finger millet's potential has not been documented previously, but today all classes of people wish to have greater awareness. Over the ages, because of its status as a poor man's cereal, it has received less attention, even in documentation. The information on finger millet has been documented as a small chapter in the books referring more on varieties and cultivation practices. In order to fill up the gap and provide adequate information, a sincere attempt has been made to bring out a book on finger millet, a valued cereal, by compiling the information with a systematic approach.

This book on finger millet covers in a methodical way, the finger millet strength, history and cultivation of varieties, nutrients, non-nutrients and the processing methods, product development and their health benefits evaluation which creates a scientific evidence for the benefits. A special and specific emphasis has been made on the starch fractions and properties; a first report on finger millet and in depth research on finger millet has been made by us in DFRL, Mysore. Additionally new coining of chapter non-nutrients has been introduced considering the functionality as well as positive and non positive properties of antinutrient factors as well as non starch polysaccharides.

This book on finger millet is a unique and first of its kind publication bringing all the relevant information together. The authors have made the maximum effort to compile the global information targeting the varieties, production and nutrients. The authors had the opportunity to bring out the first hand contribution on flavours of finger millet, starch fractions, hilly varieties, and flavorful puffed flour based convenience mixes as well as couple of products clinical study on school children.

Thus the purpose of this book is to bring out up-to-date scientific information on finger millet to the researchers to have further practical application. This book has the pride of place in the hands of food scientists, technologists, nutritionists, biochemists, agricultural scientists, cereal chemists, industrialists, as well as the consumers of the finger millet products. This book has 9 chapters on various aspects and under each chapter particular aspect has been dealt; thereby, some of the references which cover few aspects may appear again in another chapter. Keeping this in view at the end, after the 9th chapter, common bibliography has been included. The scope is immense for budding scientists since many aspects are open for further research.

ACKNOWLEDGMENTS

I am indebted to Defence Research and Development Organization, Ministry of Defence, Government of India, New Delhi and Defence Food Research Laboratory, Mysore for the support rendered during writing this valuable book. My sincere thanks to Dr.W.Selvamurthy, Chief Controller, Research & Development, Defence Research and Development Organization, New Delhi for obliging to write the FOREWORD for this book. I sincerely thank Dr.A.S.Bawa, Director, Defence Food Research Laboratory, Mysore for the support. My sincere thanks to my co-authors for literature support and timely help. I wish to place on record my sincere thanks to Smt.Pushpa U Rao who gave the computational support. My heartful thanks to my husband Sri.M.S.Ramakrishna for the constant encouragement to complete the task. My heartful thanks to my daughter Dr.Divyalakshmi. M. R and Son- in- law Sri.Chandrashekar. S. for their constant support and wishes to complete the writing of this book.

I thank the M/s Nova Publishers, New York for offering to write this book on "Finger Millet: A Valued Cereal" and bringing out this book to light.

Last but not the least, most heartfelt thanks to Almighty for showering the direction, guidance and strength to complete the work as per the schedule. I humbly submit this book to the lotus feet of the Lord.

LIST OF ABBREVIATIONS

DFRL	: Defence Research and Development Organisation
USDA	: United States Department of Agriculture'
FAO	: Food and Agriculture Organization
WHO	: World Health Organization
HTST	: High Temperature Short Time
Tg	: Glass Transition temperature
SD	: Starch Digestibility
IVPD	: In Vitro Protein Digestibility
FM	: Finger Millet
BV	: Biological Value
DC	: Digestibility Coefficient
PER	: Protein Efficiency Ratio
NPU	: Net Protein Utilization
SMP	: Skim Milk Powder
FMP	: Finger Millet Protein
RTE	: Ready To Eat
JFST	: Journal of Food Science and Technology
LWT	: Lebensmittel-Wissenschaft & Technologie

In: Finger Millet: A Valued Cereal
Editor: K. S. Premavalli

ISBN: 978-1-62081-224-2

Chapter 1

FINGER MILLET

K. S. Premavalli
Head, Food Preservation Division, Defence Food Research Laboratory,
Siddharthanagar, Mysore, Karnataka, India

INTRODUCTION

Millets are a group of small seeded species of cereal grains widely grown around the world and India is the world's leading producer of millet. It is a staple grain for much of the population and has been cultivated for thousands of years as early as 2700 BC in China. Millet runs as the sixth most important grain in the world and is a significant part of the diet in India, Africa, China, Japan and Egypt. Millet is also used as cattle feed in many of the developed countries. Sometimes millets are referred as a "Poor man's Cereal" because of relative low cost, cultivation environment as well as people when given a choice prefer other cereals such as rice and wheat. Probably, the strength of millets has not been well understood and its potential has been untapped. Millet is a superior food source in terms of dietary fibre, minerals, phytochemicals, B-vitamin series and the starch properties, as non glutinous cereal and for its physiological action. Millets are important crops in semiarid and tropical regions of the world due to their resistance to pests and diseases, short growing season and productivity under heat and drought conditions when major cereals cannot be relied upon to provide sustainable yields. Most of the millets of the world is grown in Asia, Africa and the USSR where it is largely consumed as food (Anderson and Martin, 1949). The nutrient composition is comparable to other cereals, both in major and minor nutrients. (Table 1.1 and 1.2) Rather millets are rich in minerals.

Table 1.1. Nutrient Composition of Millets and Other Cereals (Per 100 g Edible Portion; 12 Percent Moisture)

Food	Protein (g)	Fat (g)	Ash (g)	Crude fibre(g)	Carbohydrate (g)	Energy (kcal)
Rice (brown)	7.9	2.7	1.3	1.0	76.0	362
Wheat	11.6	2.0	1.6	2.0	71.0	348
Maize	9.2	4.6	1.2	2.8	73.0	358
Sorghum	10.4	3.1	1.6	2.0	70.7	329
Pearl millet	11.8	4.8	2.2	2.3	67.0	363
Finger millet	7.7	1.5	2.6	3.6	72.6	336
Foxtail millet	11.2	4.0	3.3	6.7	63.2	351
Common millet	12.5	3.5	3.1	5.2	63.8	364
Little millet	9.7	5.2	5.4	7.6	60.9	329
Barnyard millet	11.0	3.9	4.5	13.6	55.0	300
Kodo millet	9.8	3.6	3.3	5.2	66.6	353
Fonio millet	8.7	3.5	3.8	8.5	73.6	360
Teff	8.5	2.2	-	2.2	73.0	345

Sources: Hulse, Laing and Pearson. 1980; United States National Research Council / National Academy of Sciences. 1982; USDA, 1984; Encyclopedia in Food Technology and Nutrition, Vol.5, 1993.

The most common and important small grains of millet species cultivated are of six types and the details are given in Table 1.3. Amongst these, finger millet possesses high calcium, potassium, sodium and good amount of iron. Finger millet is also having relatively high quantity of dietary fibre to the tune of 19% which stands next to Kodo millet.

Table 1.2. Nutrient Composition of Millets and Other Cereals (Per 100 g Edible Portion; 12 Percent Moisture)

Food	Calcium (mg)	Iron(mg)	Thiamine (mg)	Riboflavin (mg)	Niacin (mg)
Rice(brown)	33	1.8	0.41	.04	4.3
Wheat	30	3.5	0.41	0.10	5.1
Maize	26	2.7	0.38	0.20	3.6
Sorghum	25	504	0.38	0.15	4.3
Pearl millet	42	11.0	0.38	0.21	2.8
Finger millet	350	3.9	0.42	0.19	1.1
Foxtail millet	31	2.8	0.59	0.11	3.2
Common millet	8	2.9	0.41	0.28	4.5
Little millet	17	9.3	0.30	0.09	3.2
Barnyard millet	22	18.6	0.33	0.10	4.2
Kodo millet	35	1.7	0.15	0.09	2.0
Fonio millet	30	3.4	0.30	0.10	3.0
Teff	110	90	0.50	0.10	2.0

Sources: Hulse, Laing and Pearson. 1980; United States National Research Council / National Academy of Science, 1982; USDA, 1984; Encyclopedia in Food Technology and Nutrition, Vol 5, 1993.

Besides, finger millet consumption regularly protects against the risk of cardiac problems, diabetes, gastrointestinal cancers (McKeown, 2002) and advisable as a substitute cereal for celiac disease patients (Chandrashekara and Shahidi, 2010). But millets lost its importance over the 1000 years, later started as a minor cereal. To-day, millets are being considered as strengthful cereal and it has gained its place because of its functional characteristics. In this book, finger millet is dealt in detail with reference to varieties, properties, processing, products and their benefits. Finger millet (*Eleusine coracana Gaertn*) is an important millet grown and contributes as one of the special species utilized. Its annual production was 4.5 million tonnes in 1996, but increased by 25% as per 2010 survey which reflects the development and utilization of the grain. In Africa and India, finger millet has been used as a staple food by millions. In Uganda, finger millet occupies 50% of its cereal usage.

Table 1.3. Millets and Their Composition

Varieties (Botanical Name)	Finger millet (*Eleusine coracona)*	Pearl millet *(Pennisetum glaucum*)	Proso millet *(Panicum miliaceum)*	Foxtail millet *(Setaria italica*)	Kodo millet *(Paspalum scrobiculatum)*	Barnyard millet *(Echinochl oa crusgalli)*
Parameters Quantity						
Protein, %	6-10.99	8.6-19.4	6.4-12.8	6-14	6.2-13.1	11.2-12.7
Fat, %	1-4.6	1.5-6.8	2.9-4.9	1.2-5.2	3.2-4.9	2.5-6.3
Crude fibre, %	2.6-.8	1.4-7.3	4.6-12	2.6-8.6	8.4-11	13.9-14.7
Ash, %	2.3-3.9	1.6-3.6	1.4-5	1.5-3.6	3-4.1	4.7-5
Starch, %	59	60.5	56.1	59.1	72	62
Dietary fibre, %	19.1	7.0	8.5	19.1	37.8	16.7
Calcium,mg %	0.33	0.01	0.01	0.01	0.01	-
Phosphorou s,mg%	0.24	0.35	0.15	0.31	0.32	-
Potassium, mg%	0.43	0.44	0.21	0.27	0.17	-
Sodium,mg %	0.02	0.01	0.01	0.01	0.01	-
Iron,mg%	46	74.9	33.1	32.6	7.0	-
Thiamine, mg%	0.48	0.38	0.63	0.48	0.32	-
Riboflavin, mg%	0.12	0.22	0.22	0.12	0.05	-
Niacin,mg %	1.3	2.7	1.32	3.70	0.70	-

Saldivar, (2003).

Relatively, finger millet was a low cost cereal, but dramatically rising now. In Kenya, the grain is sold at more than double the price of sorghum and maize. In India, finger millet costs ₹.15 to 18 per Kg. as against ₹.30 to 35 per Kg. for rice and wheat.Finger millet is called as Ragi in India and the common names in other countries is given in Table 1.4. Finger millet was introduced to India probably more than 3000 years ago. In Asia, the major finger millet producing country is India followed by Nepal, China,

Afghanisthan, and while in Africa, Uganda ranks first followed by Ethiopia, Tanzania and Zimbabwe (Table 1.5) and is also grown to lesser extent in other parts of Africa. Other countries which grow finger millet as a minor crop include Iran, Sri Lanka, Myanmar, Bhutan, Iraq, Pakistan and Jordan.

Table 1.4. Common Names of Finger Millet (*Eleusine coracana)*

Country	Common Name
India	Ragi
English	Finger millet, African millet, Koracan
France	Petctmil, coracan
Germany	Fingerhirse
Ethiopia	Dagussa, Tokuso, Barankiya
Kenya	Wimbi, mugimbi
Malawi	Mawere, mawe, khakwe, lipoko
Nepal	Koddo
Sudan	Tailabon, Ceyut
Tanzania	Mwimbi, mbege
Uganda	Bulo
Zambia	Kambale, lupoka, bule
Zimbabwe	Rapoka, zviyo, poho

Table 1.5. Production Profile of Finger Millet

Country	Production x 1000 tonnes
Asia	
India	2613
Nepal	122
China	79
Afghanistan	38
Africa	
Uganda	451
Ethiopia	198
Tanzania	95
Zimbabwe	46

Encyclopedia of Food Science and Nutrition, 2003.

Over the years, improvement is seen regarding promising varieties, hybrid varieties, and white coloured varieties of finger millet. However, many a time, it is considered as a poor man's crop or a famine food or birdseed. Finger millet is tasty with a mildly sweet, nut-like flavor and contains beneficial nutrients. Finger millet is rich in calcium (355 to 450 mg %), iron (10 to 19 mg %) and a good source of dietary fibre (Premavalli and Bawa, 2004a). Finger millet starch is considered as superior to wheat. Finger millet protein is having good amount of methionine, tryptophan, cystine, glutamic acid and aromatic amino acids which serve as a good source of protein for vegetarians. Millets also contain phytates, polyphenols, tannins, trypsin inhibitors and dietary fibre which were considered as antinutrients due to their chelating and enzyme inhibition activities. (Thompson, 1993) But in the present scenario, these are considered as functional factors which at specific levels impart the benefits. Finger millet can be an ideal crop in dry areas and the grain is resistant to insects, has a long storage of about 30 years. Generally, grain is ground into flour and is used for making roti, cooked balls called Ragi Mudde, dosa, weaning food and malt in India. While, flat breads, porridge, malt, alcoholic drinks in Africa. Since it is consumed as whole, fibre, minerals, phenolics, vitamins present in the seed coat are retained which are beneficial (Antony et al, 1996). As the grain is gaining its importance, several processing methods have been adopted for better strength and products development. There is an immense potential to process millet grains into value added foods and beverages in developing countries and millet products can be utilized by the developed countries because of antidisease properties. The studies on scientific evidence for the beneficial properties in humans is gaining importance but with slow pace. These aspects are dealt in detail with the up to date literature in various chapters of this Book.

Conclusion

Millets, a small seeded species of cereal grains is superior as a food source in terms of minerals, phytochemicals, dietary fibre, B vitamins and starch properties. Amongst 6 commonly used millets, finger millet called Ragi in India is most promising with high calcium, potassium, sodium and good amount of iron. Relatively high fibre and slowly digestible starch in finger millet are the positive strength for health benefits. Finger millet consumption regularly protects against the risk of cardiac problems, diabetes, gastrointestinal cancer and advisable as a substitute cereal for celiac disease patients.

In: Finger Millet: A Valued Cereal
Editor: K. S. Premavalli
ISBN: 978-1-62081-224-2

Chapter 2

HISTORY, CULTIVATION AND VARIETIES OF FINGER MILLET (*ELEUSINE CORACANA GAERTN*)

K. S. Premavalli and Y. S. Satyanarayana Swamy
Food Preservation Division, Defence Food Research Laboratory,
Siddharthanagar, Mysore, Karnataka, India

INTRODUCTION

Finger millet called Ragi is a widely cultivated crop of the tropical and subtropical regions of the world. Finger millet is considered as dryland crop, Famine crop, but certainly promising millet because of nutritive properties. It can be grown under rainfed conditions and under irrigation, as well as in poor soils of hilly area to rich soil in the plains. The global view shows more common cultivation from sea level to altitudes of 3000 m. Relatively, it occupies a lower position among food crops, still remain as staple food for some population.

1. HISTORY

Finger millet as Ragi is mentioned in India by the ancient Sanskrit writers referring as 'Rajika'. Eleusine coracana is originated from the wild species

Eleusine indica. It was under cultivation before Aryans reached the Indian peninsula. Decondelle (1886) is of the opinion that finger millet is originated in India and then spread to Arabia and Africa, nearly 3000 years ago (Gururaj and Krishna, 1998). But as per Sundarraj and Thulsidas (1993), Vavilov in 1951 proposed that E. coracana originated in Ethiopia. While Mehra in 1963 opined that finger millet is considered to be African origin and an early migration to India. The name *Eleusine* is a generic one and is given after Eleusine, the Greek goddess of cereals. Scientifically, prolamine, the major protein fraction in finger millet called eleusinin may also be the reason for the name eleusine. The common name, finger millet is derived from its branching of the panicle like fingers of the hand. Though the origin of finger millet is debated, the regular cultivation and utilization is in Asia and Africa.

a)

b)

Figure 2.1. Finger Millet Matured Plant of Indian Variety.

2. Botonical Description

Finger millet is an annual crop growing to 2 to 4 feet in height with a firm and fibrous root system which aids in absorbing moisture efficiently from the soil. The stem of the plant is compressed with nodes and the leaves are linear with distinct mid rib, ligule and a fringe of hairs. The tillers which are generally well formed, bear at the end of the short slender culms which consists of a whorl of finger like spikes normally 2 to 8 in number. The spikelets are found closely around the finger may be crowded with overlapping rows on the outer side of the spike. The flowering on a spikelet may range to 4 to 5 which may take 6 to 8 days for flowering. The fertilization and maturity of the seeds covers 90 to 130 days leading to brownish colour shades of seeds. The matured plant of Indian variety grown in Karnataka state is shown in Figure 2.1.

3. Cultivation Practices

The natural condition of land used for cropping differs from place to place and the cropping duration varies. Finger millet grows well on drained loam or clay loam soils, heavy black cotton soils, stony soils and tolerates salinity better than other cereals. Planting time depends upon the rains and the time taken to prepare the seed bed. The planting time is November to January in Tanzania, March to April in Kenya, December to February in Uganda, April to May in India for the rainfed crops. For the dry land crops, irrigated crops, the planting time change accordingly. Finger millet is grown successfully from sea level to an altitude of 2500 meters on hilly regions as well as plains. In India it grows up to 2100 m above sea level (asl), 2400 m asl in Uganda and 2500 m asl in Nepal. The annual average rainfall of 50 to 100 cms is enough for good cultivation. Sowing of seeds is done by broadcast after digging or ploughing and the seeds requirement is about 5 to 6 Kgs per hectare. Systematised sowing by making ridges and rows and seeding is not generally followed in finger millet cultivation. Though line sowing ensures better germination, cuts down seed requirement and facilitates intercultural operations, the method is in lesser practice. The method of on line sowing is discussed by Chidda Singh et al (2003). In rainy areas, transplantation of crops will be promising. About 4 Kg seeds can be used for raising nursery and transplanted on lining. Stilled water after rains is not recommended and removal of excess water is essential.

Table 2.1. Recommended Fertilizer Doses for Finger Millet in Different States

State	Fertilizers dose N:P:K (Kg/ha)	
	Rainfed	Irrigated
Andra Pradesh	40:20:20	90:30:20
Bihar	40:20:20	60:30:20
Gujarat	80:40:0	-
Himachal Pradesh	40:20:0	-
Karnataka	50:40:25	100:50:50
Maharashtra	25:20:0	50:25:0
Madhya Pradesh	40:40:0	-
Orissa	40:20:20	60:20:20
Tamil Nadu	60:30:30	90:45:45
Uttar Pradesh	60:30:0	-
Uttaranchal	40:20:0	-

Chidda Singh et al, 2003.

Table 2.2. Profile of Intercropping Systems in Finger Millet Followed in Various Places

Countries	Intercrops
India South North	 Groundnut, Tobacco, Sugarcane, Potato-Maize Potato-finger millet Chickpea, Barley, Mustard, Linseed, Tobacco
Tanzania	Maize, Sorghum, Cassava, Pumpkin, Sunflower, Sesame
Uganda	Maize, Sorghum, Peas,
Nepal	Maize
Kenya	Maize, Sorghum, Cassava, Cotton
Ceylon	Sesame

Compiled from different sources.

The crop yield will be better with transplantation. Generally, fertilizer is not used in most of the countries. In recent years, to improve the soil strength, low levels of manures and fertilizers are suggested. Relatively, fertilizers quantity are higher for irrigated crop as compared to rainfed crops (Table 2.1). The irrigated crop of finger millet is sown in more than one season in South India. The maturity period ranges from 90-140 days depending on the tract and variety. Blast disease and pest infestation is more common in finger millet plants. Blast of finger millet caused by Pyricularea grisae can result in losses greater than 50-80%. Timely treatment of plants and eradication of infested plants can be effective control measures (Vishwanath and Seetharam, 1989; Gururaj and Krishna, 1998). Weed control is one of the most difficult problems in finger millet production. Weeding is normally followed by hand or big hoes, knives and two to three weedings will be enough to control the weeds. Intercropping is a common practice with finger millet and generally intercropped with maize, legumes, etc., but it differs with place and type of cultivation practices followed. The profile of intercropping system followed in various places is given in Table 2.2. Intercropping with legumes adds to the nitrogen balance of soil. However, Hegde and Linge Gowda (1989) are of the opinion that traditional intercropping with a minor legume such as field bean may not be remunerative in irrigated conditions, although highly suited to dryland marginal conditions. The harvesting is generally done soon after ripening in two stages. The earheads are harvested with sickles or curved knife and straw is cut close to the ground. Earheads are heaped for 3 to 4 days to cure and then thrashed with hand or bullocks. Alternatively, whole plant with earhead is cut, heaped and then thrashed. However, the harvesting and cleaning of finger millet is a laborious process. Further, storage is done in large baskets with the walls made firm by plastering wet soil.

4. FINGER MILLET VARIETIES AND PRODUCTION

Earlier years, local varieties were grown which had a low yield of 500-1000 Kgs per hectare. Research has been carried out for improving the varieties with better yield. During the 80's, significant progress has been made to identify specific characters associated with higher productivity under moisture stress using varied physiological approaches (Udaya Kumar et al, 1989). Indian scenario on the production and productivity of finger millet can be seen from Table 2.3. The crop yield has slowly improved from 678 to 1420 Kgs per hectare from 1951-2000. Among the various states in India, Karnataka rates high in production of finger millet (Table 2.4). In Karnataka, Hamsa, Dibya Sinha, Sarada are the earlier varieties achieved by mutation breeding. Hybridisation of indigenous germplasm have yielded derivatives such as Udaya, K7, Purna, Annapurna, Cauvery, Shakti, HPB76 showed intermediate productivity range (Harinarayana, 1989). Pure line selections have contributed the maximum number of ragi varieties. Nirmal and Birsa Mandya are early varieties with 2 to 2.5 tonnes per hectare yield, while, Simhadri and PE176 are mid late varieties with the same yield potential. Late varieties Ratnagiri, Nilachal, PE6110 and Paiyar1 showed between 2 to 2.25 tonnes per hectare. Research progress has evolved a number of high yielding varieties which meets the specific requirements of different regions i.e. dry areas, saline and sodic soils, early and late planting, blast resistant, coastal and rainfall regions. The comparison of local and improved varieties have revealed the potential of higher yield of improved varieties (Table 2.5).

Table 2.3. Area, Production and Productivity of Finger millet (1951-2005)

Year	Area (Xooooha)	Production (Tonnes)	Productivity (Kg/ha)
1951-55	2246	1520	678
1956-60	2414	1874	778
1961-65	2515	1991	791
1966-70	2465	1721	703
1971-75	2409	1975	820
1976-80	2609	2726	1042
1981-85	2499	2593	1036
1986-90	2346	2444	1084
1991-95	2015	2542	1267
1996-2000	1826	2586	1420
2001-2005	1630	2098	1276

Seetharam et al, 2001.

Table 2.4. Production of Finger Millet in Different States of India (1998-99)

States	Area, %	Production, %
Karnataka	57	64
Maharashtra	9	6
Tamil Nadu	9	12
Uttar Pradesh	9	8
Andra Pradesh	6	5
Orissa	4	2
Bihar	4	2
Gujarat	1	1
Madhya Pradesh	1	<1

Seetharam et al, 2001.

Table 2.5. Finger Millet Yield in Local VS Improved Variety in India

States	Local variety Kg/ha	Improved variety Kg/ha
Uttaranchal	1496	VL149 : 2365
Jharkhand	1774	A404 : 2380
Orissa	1950	PR202 : 2734 VL149 : 2734
Karnataka	1214	HR911 : 2173
Maharashtra	820	HR374 : 1010

Seetharam et al, 2001.

Hybridization of indigenous and African exotic germplasm has yielded good derivatives such as Indaf 9, Indaf 1, Indaf 5, Indaf 7, HR911 with their productivity at 2.5 tonnes per hectare. The crop improvement and generation of new varieties is a continuous pro8gramme, however, in India the crop yield of hybridized variety is not exotic. The development of new varieties of finger millet is equally a great attempt being made in Africa. In Uganda, in a study during 1985 on comparison of 29 popular local varieties and 71 exotic introductions, indicate that local farmer varieties yield better than exotic varieties and thus focused on the improvement of existing varieties (Zake and Khizzah, 1989). However, the yield reported ranged from 5.2 to 21.8 quintals per hectare. A field survey in Tanzania on local finger millet varieties showed that high altitude types are quite productive yielding up to 50 quintals per hectare. The research programme on large collections of local varieties with reference to their botanical characteristics, chemical composition, yield potential and agronomic characteristics have shown some 22 promising varieties with an yield of 19.4 to 44.4 quintals per hectare (Mwambene, 1989). In Nepal, about 350 local land races and 150 exotic introductions mainly from East Africa when tested showed variations in terms of maturity, plant type, pigmentation and reaction to diseases (Deepman Sakya, 1989). In India, the production of finger millet has improved with the development of Indaf varieties. The crop variety development from 1940 to 2005 referring to the year of release, growth duration and yield is presented in Table 2.6. It can be observed that upto 1960's, the yield was less, limiting to 1300 Kgs per hectare, irrespective of single local variety, or mutant. In 1970, 'White ragi' has been reported which gave an yield of 4500 Kgs per hectare. Later the hybridization of germplasms of both Indian or Indian and African have yielded 1800 Kgs to 5000 Kgs per hectare. The same trend followed upto 2005. But the growth maturity duration did not reflect a clear cut pattern.

Table 2.6. Profile of Finger Millet Cultivars

Sl.No.	Name of the variety	Hybrid/Common name	Year of release	Duration (Days)	Yield (Kg/ha)
1	AKP1	Arakka Patti strain	1940	90	900
2	AKP2	Arakka Patti strain	1940	85	900
3	CO 1	EC 593	1942	120	1135
4	CO 2	EC 3517 Multi ragi	1942	110	929
5	CO 3	Mutant foam Co1	1942	110	910
6	PL R1	Permuragi	1942	110	910
7	CO 4	Local ragi of Palladam	1942	140	1200
8	K 1	Koril patti	1948	135	910
9	CO 7	Local ragi	1953	100	1364
10	CO 5	Economics set of millet branding st	1953	115	1364
11	CO 6	Sel. From ES 15GUxE52985	1953	120	773
12	K 2	Karnom surattai rai	1955	135	1365
13	CO 8	Nellore	1963	85	1000
14	CO 9	White ragi	1970	100	4500
15	CO 10	Marna ragi	1973	85	3500
16	Indaf 5	Kaveri + I.E. 927	1977	115	2500-3000
17	CO 11	PLS from MS 2584	1978	95	4000
18	Indaf7	Annapoorna + IE 927	1981	120	4500-5000
19	Indaf8	IE 929 + Grass crop	1982	120	3500-4000
20	CO12	Sel. Foam PR 722 of Reddapuram	1985	110-120	3928
21	HR911	UAS 1 + IE 927	1985	118	3500-4000
22	PR 202	A.P. variety	1985	118	3500-3000
23	Indaf 9	K.I + IE 980	1985	105	2500-2000
24	Piazas1	Pure line	1986	115-120	3125
25	CO13	CO 7 x Taliko 7	1988	105	3600
26	Trichy1	Sel from HR 574	1989	100	4011
27	PES400	1305 x Co13 Pure line selution	1989	98-102	1800-2000
28	TRY1	HR-37←Selectron from	1989	100-105	2000-2500
29	VL124	Sel from Local Ger	1989	95-100	2000-2500
30	RAU8	BR407 x Romchi local	1989	105-110	2200-2500
31	MR1	Hamsa + IE 927	1990	125	4000-4500
32	Indaf15	IE 67 + IE 927	1991	130	4500-5000
33	VL149	VL 204 x IE 882	1991	98-102	2000-2500

Sl.No.	Name of the variety	Hybrid/Common name	Year of release	Duration (Days)	Yield (Kg/ha)
34	MR1	Harsh x IE 927	1993	125-130	2500-3000
35	A 404	Sel. From Local G	1993	110-115	2200-2500
36	GN3	Kon 13 x GN2	1993	130-135	2200-2500
37	Km65	Sel from EX etic Ger	1994	98-102	1800-2100
38	Ganthami	PR 202 x 422	1994	115-120	2800-3000
39	VL-146	Sel. From Germ plasm	1995	95-110	2500-2800
40	Padmavathi	Sel. From Tirupathi local	1995	100-105	2500-3000
41	MR2	PR 202 x IE 927	1995	125-130	3500-4000
42	Bm2	Selectran	1995	105-110	2400-2600
43	GPG28	Indaf 5 + Indaf 9	1996	110-115	3500-4000
44	VR708	Sel. From local	1997	90-95	2500-3000
45	PR 230 (Mosuti)	Pure sinc selectron	1998	90-100	2500-3000
46	BM 91	Mutant from Budha	1999	103-105	2500-3500
47	GPU45	Indaf 5 + Indaf 7	1999	105	2500-3000
48	GPU26	15x19xIE 1012	2000	100-105	3000-3500
49	CO(Ra)14	A desirative involving malavi	2004	105-110	2892
50	MR6	RW 1 + ROH 2	2004	120	4500-5000
51	GPU48	Indaf 5 + Indaf 7	2005	100	2500-3000

Ravishankar and Prakash, (2007).

However, 85-135 days is the period followed. When the cropping system, sowing of seeds, irrigated crop or rainfed crop and weed control are considered, the equations for duration and yield may fit and these factors have been well discussed (Gururaj and Krishna, 1998; Deepman sakya, 1989 and Chidda singh et al, 2003). But some of the varieties are more advantageous in terms of early maturity and resistance to blast. The shift in monsoon and occurance of blast in the recent years necessitated evolving short or medium duration varieties with blast resistance, which resulted in GPU varieties. GPU28 yields 40 to 45 quintals per hectare under irrigation and 30 to 35 quintals per hectare under rainfed conditions. GPU26 is an alternative to Indaf 9 yielding 30 to 35 quintals per hectare under rainfed conditions. While GPU 48 is short duration variety, but blast tolerant, gives 37 quintals per hectare under rainfed conditions. Thus there are so many replaceable, improved varieties. But some of the local varieties such as MR1, MR2, MR6 i.e. Mandya Ragi still remain promising. The list of varieties grown in different states of India is given in Table 2.7.

Table 2.7. Varieties of Finger Millet Grown in Different States of India

Sl. No.	Name of the State	Name of the variety		
1.	Andra Pradesh	AKP 2, Sharada V 2M 1 Gouthami Kalyani	Godavari Kalyani Simhadri Padmavathi Sapthagiri	Suraj VR708 HR374, PR2 AKP1 Ratnagiri
2.	Bihar	BR2 BR407 RAG8 HR374	RAU5 A404 Bm2 VL149	
3.	Gujarat	Gujarat Nagli 1 Gujarat nagli 2	GN3 BR407 VL149	
4.	Himachal Pradesh	VL 204 VL149 VL124		
5.	Karnataka	HR 374 Indaf 9 Indaf 1 Indaf 11 MR 2	Indaf 3 Indaf 8 HR 911 HR 919 GPU 28	PR 202 Indaf 5 Indaf 7 MR 1, Ham HR 374
6.	Maharashtra	B 11, HR 374 A 16 E 11	28 1, BR 407 50 1 VL 149	RAH 8 BM 9 1 PES 176
7.	Madhya Pradesh	JNR 852 IE 28 ECU 840 VL 149	PR 202 RAU 8 BM 9 1 JNR 1008	JNR 981 JNR 852 BR 417
8.	Orissa	Dibyasingha B 4 10 56 PR 202	VL 149 Bhudei 04AJ 2 OUAT 2	VR 708 BM 9 1
9.	Tamil Nadu	CO7 CO10 K5 K6	Indaf-5 PR 202 CO 11 K 7	CO12 CO13 Paiyur 1 TRY 1
10.	Uttar Pradesh	Nirmal T 306, PES 110	PES 176 VL 204	KM 13 KM 65
11.	Uttaranchal	VL 124, VL 149 VL 146	VL 204	PES 400

Note: (1) VL 149 is a blast tolerant, early and high yielding variety.
(2) VL124 is superior in both seeds and fodder yield.
(3) OUAT2 is a white seeded variety.
(4) TRY 1 is tolerant to salinity.
(5) Paiyur 1 possess drought tolerance.

Table 2.8. Recommended Varieties of Finger Millet for Different States

State	Varieties
Andra Pradesh	SurajVR 708 (Champavathi) Godavari, Ratnagiri, Gautami Padmavathi, Kalyani, Sapthagiri, RR 230 (Maruthi)
Bihar	RAU, 8, VL 149
Chattisgarh	PR 202, RAU 8, BM 9 1
Gujarat	Gujarat Nagli 2, VL 149
Himachal Pradesh	VL 124, VL 149
Jarkhand	A 404, Birsa Marua 2 (BM 2), VL 149
Karnataka	Indaf 9, GPU 26 Indaf 8, HR 911, PR 202, MR 1 PR 202 and GPU 28 Indaf 7 Indaf 5
Maharashtra	VL 149, RAU 8, BM 9 1
Madhya Pradesh	VL 149, PR 202, RAU 8, BM 9 1
Orissa	PR 202, VL 149, VR 708, OUAT 2, BM 9 1
Tamil Nadu	Co 11, Co 12, Co 13, K7, Paiyur 1, TRY 1Co 7, Indaf 5 VR 708
Uttar Pradesh	KM 13, KM 65, PES 110
Uttaranchal	VL 124, VL 204, PES 400, VL 149

Seetharam et al, 2001.

Table 2.9. Promising Finger Millet Varieties for Different States

State	Varieties
Andra Pradesh	Padmavathi, Gautami, HR 374, AKP 7
Bihar	RAU 5, RAU 8, A 404, BM 2
Gujarat	GN 3, BR 407
Karnataka	MR 1, MR 2, GPU 28, HR 374, Hamsa
Madhya Pradesh	JNR 852, JNR 1008, BR 407, PR 202
Maharashtra	PES 176, HR 374, BR 407
Tamil Nadu	CO 13, TRY 1
Uttaranchal	PES 400, VL 124, VL 149, KM 65, VL 146, PES 176

Chidda Singh et al, 2003.

Table 2.10. Salient Features of Some of the Important Finger Millet Varieties

Varieties	Maturity (days)	Average yield (q/ha)	Special features
PES176	102-105	20-22	The average plant height is 88 cm. Seeds are of brown colour. It is resistant to blast.
PES400	98-102	18-20	It is early maturing variety. It is resistant to blast.
VL146	95-100	25-28	It is resistant to blast.
VL149	98-102	20-25	Blast resistance, wide adaption and earliness
VL124	95-100	20-25	Earliness, good for seed as well as fodder
CO13	110-120	25-30	For Tamil Nadu State
TRY1	100-105	20-25	For Tamil Nadu State. It is tolerant to salinity.
RAU8	105-110	22-25	Blast resistance, wide adaption and earliness.
MR1	125-130	25-30	For Karnataka. It has high yield potential. It is resistant to major diseases.
MR2	125-130	35-40	It is resistant to major diseases.
A404	110-115	22-25	For Bihar. Blast resistance.
BM2	110-115	24-26	For Bihar. It is resistant to blast.
GN3	130-135	22-25	For Gujarat.
GPU28	110-115	35-40	For Karnataka. It is highly resistant to blast.
HR374	110-115	20-22	For Karnataka and Maharashtra.
KM65	98-102	18-20	For hills of Uttaranchal.
Gautami	115-120	28-30	It is tolerant to blast. It is suitable to grow in Andhra Pradesh.
Padmavathi	100-105	25-30	For Andhra Pradesh. It is resistant to blast.
JNR852	100-112	22-25	For Madhya Pradesh. It is resistant to blast.

Chidda Singh et al, 2003.

But the advancement in cultivation of many varieties of finger millet under different weather conditions has led to the identification of varieties for various regions. The recommended varieties of finger millet along with seasons for different states are presented in Table 2.8. More number of varieties are suitable for Andra Pradesh, Karnataka and Tamil Nadu, southern states of India, which reflects on the suitable cultivation conditions as well as increased consumption. Further in 2003, Chidda singh et al have suggested most promising varieties for 8 states (Table 2.9). The varieties not found in earlier report of 2001 (Seetharam et al) have been covered in his report which may reflect the continuing developmental programmes for improvement of cultivation practices and varieties.

The salient features of some of the important finger millet varieties are shown in Table 2.10. It is interesting to observe that many of the local varieties are found and mostly their properties thereby better yield have more impact for their importance. Similar phase of application of improvement of yield was attempted in African countries and a brief of the varieties in Nepal, Tanzania and Zimbabwe is given in Table 2.11. Minimum 9 to10 varieties are being cultivated and utilized by the population as a staple cereal. In Uganda, Serere1, GuluE and P224 are the higher yielding varieties, while, in Kenya P283, P224 and P221 were found to be most promising.

5. Varieties vs Physico-Chemical Characteristics

Among the millets, finger millet is more different in colour of grains and this property many times is the limiting factor for acceptance of finger millet products. The finger millet seed coat contains coloured pigments tightly bound with the soft and friable endosperm. The characteristic brown colour of the grain has been called as 'ragi brown'. The factors responsible for colour of the grain has been discussed by Sundararaj and Thulsidas (1993). Two factors B1 and B2 are capable of producing the colour either alone or in combination. A third factor S in association with the B factor produces purple plant pigmentation and a factor D combination with B factor deepens the effect. Thus finger millet colour shades are seen from pale brown to blackish brown. The list of the varieties with the grain colour is given in Table 2.12. Chemically tannins are the compounds which impart brown colour and the varieties devoid or reduced tannin varieties will be pale white finger millet called 'Majjige ragi" in India.

Table 2.11. Finger Millet Varieties Grown in Different Countries

Sl.No.	Name of the country	Varieties	Yield (Kg/ha)
1.	Nepal	NE 1703- 34	1789
		NE 6401 - 26	1547
		NE 1104 - 13	1261
		NE 1001 - 1	1415
		NE 1102 - 12	1162
		NE 1304 - 1	1059
		Dalle -1	1424
		NE 3801 - 2	1396
		Okhale -1	1217
2.	Tanzania	Mbeya local	4233
		Sumbawnga	3256
		Engenyi	2437
		Serere local	2326
		P 283	2293
		P 224	2327
		Gulu E	1924
		Rombo local	1630
		Edling	2286
3.	Zimbabwe	TGR -367	3300
		TGR -54	2790
		TGR -164	2720
		TGR -316A	2630
		TGR -47	2620
		TGR -402	2620
		TGR -327	2560
		TGR -295	2520
		TGR -126	2500
		TGR -209B	2480
		TGR -389	2470
		TGR -316B	2470
		TGR -72	2440
		TGR -294	2360
		TGR -362	2350
		TGR -35	2280
		TGR -303	2270
		TGR -84	2250
		TGR -108	2250

Sl.No.	Name of the country	Varieties	Yield (Kg/ha)
		TGR -10	2210
		TGR -127	2200
		TGR -113	2190
		TGR -304	2170
		TGR -316	2110
		TGR -73	1960
		TGR -83	1920
		TGR -42	1910
		TGR -36	1880
		TGR -22	1690
		TGR -34	1590

Seetharam et al, 2001.

However, the properties other than tannins of these varieties have an influence on the products and these aspects are dealt in chapters on nuitrient and non nutrients. Sankara et al (1998) have reported the cultivation of 36 varieties with varied seed coat colours. They found that white seeded genotype showed higher protein content while brown seeded genotype exhibited a wide range of variations. Sieucela et al (2007) analysed grains of 22 finger millet type from Africa to understand the localization of tannin as well as its influence on other properties. Eighteen were brown varieties while 4 were creamy white. All the kernels except light type stained black with the bleach test and showed higher tannin content and antioxidant activity, while light coloured grains had much lower tannins and lower total phenols. Light microscopy indicated that the kernel stained black had a dark coloured test layer inferring that tannins were located in that layer, and is the first report made on thc blcach tcst for dctcction of tannins in finger millet types.

Premavalli et al (2006) have studied the finger millet kernels properties in terms of weight of kernels and bulk density. The colour of the grains studied is presented in Figure 2.2 The study involved both hill grown and base varieties. The weight of 100 kernels ranged from 0.29 to 0.39g in base varieties, while 0.23 to 0.26g in hilly varieties which reveals that the grains may be small in hilly regions because of lower temperatures of cultivation. The bulk density in terms of grams per litre ranged from 770 to 880 in base varieties and 760 to 780 in hilly varieties except the local white variety WRC112 had 770 and Indaf11 had 720. It is clearly brought out that the hilly varieties and white variety have low bulk density. The influence of varieties on the nutrient composition, antinutrient factors, antioxidant activity etc. have been studied.

Table 2.12. List of Finger Millet Varieties VS Grain Colour

Name of the country	Sl. No.	Varieties	Grain Colour	References
Africa	1.	IE 927	Brown	Geeta et al,1977
	2.	IE 929	Brown	
	3.	IE 974	Brown	
	4.	IE 976	Brown	
	5.	IE 978	Brown	
	6.	IE 979	Brown	
	7.	IE 1029	Brown	
	8.	IE 1038	Brown	
	9.	IE 1039	Brown	
	10.	IE 1065	Brown	
India	1.	Purna	Brown	Siucela etal,
	2.	HPB 20-25	Brown	2007.
	3.	HPB 1-8	Brown	
	4.	HPB 7-16	Brown	
	5.	HPB 23-26	Brown	
	6.	EC 4840	Brown	
	7.	IE 246	Dark Brown	
	8.	IE 328	Dark Brown	Sankara etal,
	9.	IE 497	Dark Brown	1998
	10.	IE 860	Dark Brown	
	11.	IE 121	Brown	
	12.	IE 395	Brown	
	13.	VL146	Brown	
	14.	VL149	Pale Brown	
	15.	VL204	Dark Brown	
	16.	Indaf5	Dusty Brown	
	17.	Indaf8	Brown	
	18.	Indaf9	Pale Brown	
	19.	MR1	Brown	
	20.	HR911	Dark Brown	
	21.	GPU28	Dark Brown	
	22.	Local	Dark Brown	
	23.	Hamsa	White	
	24.	HPW	White	
	25.	HPW274	White	
	26.	HPW8384	White	
	27.	ECW 955	White	
	28.	HX 799	White	
	29.	Indaf11	Creamy White	
	30.	WRC112	Creamy White	

Name of the country	Sl. No.	Varieties	Grain Colour	References
	31	CO10	Dark Red	Ravindran,
	32	KM1	Dark Red	1991
	33	MI 302	Very Dark	
	34	WR9	Red	
	35	WR-5	White	
	36	Indaf11	White	
	37	PR202	White	UdayaSekharRao
	38	HR374	Brown	and Deosthale,
	39	JNR 1008	Light Brown	1988
	40	PES 8	Brown	
	41	PES176	Brown	
	42	JNR852	Brown	
	43	HR919	Brown	
	44	PES110	Brown	
	45	RAU8	Light Brown	
	46	BR407	Dark Brown	
			Dark Brown	

Compiled from different sources.

In order to explore the possibility of using the best suitable variety, such research attempts are of high value. Ravindran (1992) has reported on 3 varieties of finger millet regarding proteins, its digestibility, amino acid composition and proteinase inhibhitors. Millet proteins were deficient in lysine but the other essential amino acids were found at adequate levels and better balance as compared to other millets. The proximate composition and minerals were analysed in 10 varieties (Barbeau and Hilu, 1993) comprising of both 2 wild and 8 domesticated varieties. The variations between the two types was observed. The wild proginator showed significantly higher level of calcium, iron, protcin and lysine levels than 2 domesticated varieties. The study indicated that the better, more promising nutritional strength can be obtained by cross breeding of wild and domestic cultivars.

The morphological and nutritional qualities of 15 varieties have been reported by Sreedharamurthy et al (1997). Amongst these, 12 varieties showed a large variation in size of the grains except Indaf 8 with small size. The fractions separation according to size revealed that first two fractions covered the major portion of sample. Dehusking in a Mengetti cone polisher showed that husk content was lowest for Poorna variety and highest for MR2 variety. Antony and Chandra (1999) have reported changes in white, a non tannin variety and brown variety with regard to in vitro protein digestibility and mineral availability.

a)

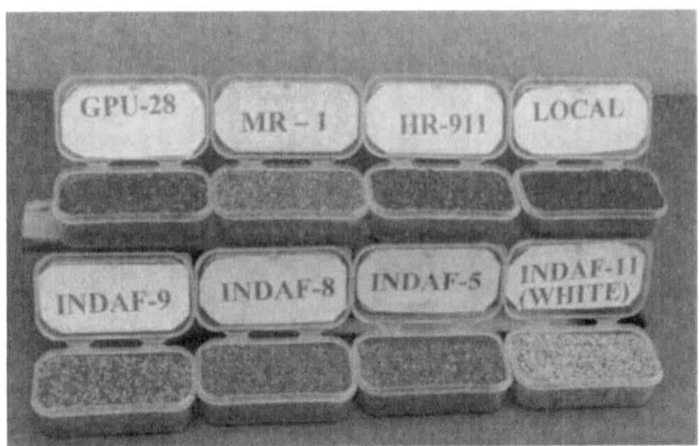

b)

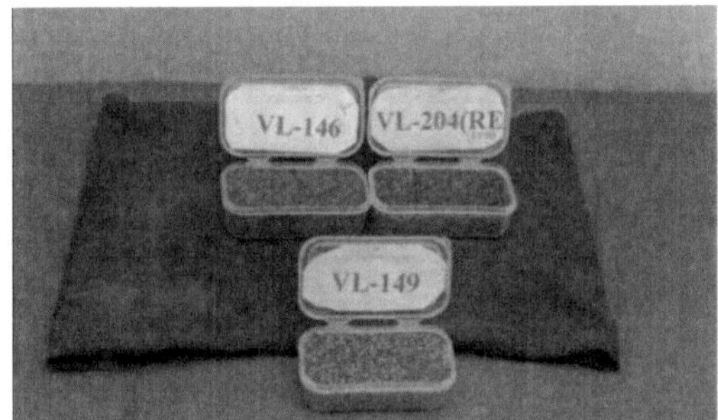

Base Varieties: GPU 28, MR 1, HR 911, Local, INDAF 9, Indaf 8, Indaf 5, Indaf 11 (White).
Hilly Varieties: VL 146, VL 149, VL 204.

Figure2.2. Finger Millet Color Profile of Indian varieties.

Thus it is very clear that the varieties differ regard to their place of cultivation, type of cultivation and have a greater impact on the properties which inturn influence the technological application for development of products, thereby the availability of nutrients and bioaccessibility of minor nutrients are affected. During 2007, Ministry of Public Distribution System, Government of India have brought out the specifications for finger millet with reference to moisture and refractions of grains (Table 2.13).

Table 2.13. Uniform Specification for Ragi (Marketing Season 2006-2007)

The Ragi shall be dried and matured grains *Eleusine coracana.* It shall have uniform size, shape and colour. It shall be in sound merchantable condition and also conforming to PFA standards.

Ragi shall be sweet, hard, clean, wholesome and free from moulds, weevils, obnoxious smell, *Argemone Mexicana* and *Lathyrus sativus* (Khesari) in any form, colouring matter, admixture of deleterious substances and all other impurities except to the extent indicated in the schedule below:

Schedule of Specification

S.No.	Refractions	Maximum Limits (%)
1.	Foreign matter *	1.0
2.	Other foodgrains	1.0
3.	Damaged grains	1.0
4.	Slightly damaged grains	2.0
5.	Moisture content	12.0

* Not more than 0.25% by weight shall be mineral matter and not more than 0.10% by weight shall be impurities of animal origin.

Ministry of Public Distribution System, Govt. of India.

N.B.

[1] The definition of the above refractions and method of analysis are to be followed as given in Bureau of Indian Standard 'Method of analysis for foodgrains' Nos. IS: 4333 (Part-I): 1996 and IS: 4333 (Part-II): 2002 and "Terminology for foodgrains" IS: 2813-1995 as amended from time to time.

[2] The method of sampling is to be followed as given in Bureau of Indian Standard 'Method of sampling of cereals and pulses' No. IS: 14818-2000 as amended from time to time.

[3] Within the overall limit of 1.0% for foreign matter, the poisonous seeds shall not exceed 0.5% of which Dhatura and Akara seeds (Vicia species) not to exceed 0.025% and 0.2% respectively.

[4] Kernels with husk will not be treated as unsound grains. During physical analysis the husk will be removed and treated as organic foreign matter.

CONCLUSION

Finger millet called Ragi, widely cultivated crop with the botanical name *Eleusine coracana* obtained the name after *Eleusin,* the Greek Goddess of cereals, but scientifically because of prolamine the major protein fraction called eleusinin. Since the branching of finger millet panicle is like fingers of the hand, finger name is figured. Finger millet is an annual crop with short period of cultivation covering 90 to 130 days, grows upto 3500m asl with an average rainfall of 50 to 100 cms and main cultivation is in Asia and Africa.

In India, Karnataka rates high in production of finger millet among the various states. Crop variety development from 1940 to 2005 showed a gradational improvement both in number of varieties and the yield per hectare. More than 50 varieties with the yield from 1300 to 5000 kgs per hectare are being used for cultivation in various states. Similar improvement is seen in African Countries. However, 9 to 10 best varieties are identified for cultivation and utilisation by the population. Among the millets, finger millet has the characteristic colour range from reddish brown to brown to dark brown with dull white as white variety and the colour is due to tannins. Over the decades, improved agronomical practices led to the development of a large number of varieties from normal to hybrid to noncoloured with maximum yield of 5000Kgs per hectare and their short growing season, resistance to pests and diseases, good productivity under heat and drought conditions has made it all the more promising cereal.

In: Finger Millet: A Valued Cereal
Editor: K. S. Premavalli
ISBN: 978-1-62081-224-2

Chapter 3

Nutrients of Finger Millet

K. S. Premavalli and D. D. Wadikar
Food Preservation Division, Defence Food Research Laboratory,
Siddharthanagar, Mysore, Karnataka, India

Introduction

Finger millet (*Eleusine coracana*) a minor-cereal, is nutritionally rich and functional staple food used by a part of African and Asian population. Among the cereals, it is an excellent source of minerals especially calcium and iron. It is an important cereal nutritionally, physiologically, medicinally and functionally, which directly or indirectly reflects on the structural, physicochemical characteristics of the grain and its chemistry. With the available literature, compositionally finger millet has carbohydrate 73 to 82%, protein 4 to 8%, calcium 200 to 450mg%, iron 5 to 15mg%, B-vitamins 0.4 to 4 mg% and crude fibre 3 to 12% and seven essential amino acids. In general carbohydrates, fat and protein are present in higher quantities and are responsible in foods for providing energy needs and body building. Apart from these, each group has a specific role to play in the human physiology. Carbohydrates though provide energy immediately, but the type of the carbohydrate plays a significant role for their action. Lipid is a rich source of energy, and storage depot but essential fatty acids have a greater role on brain functioning, cardiac health, control of blood serum lipid profile, skin health etc. Protein though take part in tissues formation inturn cells inturn body development, the type of protein, the quantum of protein and essential amino acids present have a greater impact on growth and maintenance of good health. Though the physiological actions based on metabolism are specific to each, their role is based on the type of the food, relative nutrition and their synergistic action. Thus it is essential to understand the composition of any food material to have an insight into the applications. Therefore in this chapter the major nutrients in terms of carbohydrate, protein, lipid of finger millet and minor nutrients in terms of minerals and vitamins are dealt with.

Generally, nutrient compositions of cereals are studied to know the potential sources. Likewise, finger millet varieties have also been studied for their nutrient composition and many a time compared with other millets and cereals. Since it will be more relevant to discuss on finger millet, the literature mostly covering the finger millet composition has been cited. The major nutrients composition of various cultivars is presented in Table 3.1 where the total carbohydrates, protein and lipid contents in various varieties have been reported. Finger millet is low in lipid, moderate in protein and normal in the levels of carbohydrates.

1. CARBOHYDRATES

Finger millet like any other cereal is a source of dietary carbohydrates which comprise of starch as the main constituent with non-starchy polysaccharide at levels of 15-20% as unavailable carbohydrates. The carbohydrate profile of 76 varieties of finger millet has been reported as given in Table 3.1 and carbohydrates vary from 61 to 88.9% and the wide range may refer to the varietal differences, and weather variations of cultivation. The fingermillet varieties studied earlier to 70's for compositional characteristics have been reviewed by Hulse et al (1980). According to Kadkol and Swaminathan (1954) 8 varieties had shown an average of 82.4% carbohydrates. In Philippines two varieties, Solpico and Yamboo (1966) have reported 73.5% carbohydrates. Subramanyan et al (1955) have reported 66.2% starch, 1.2% reducing sugars, 6.8% hemicelluloses and other carbohydrates in H22 variety. Pore and Magar (1977) have reported on the nutritive value of 36 hybrid varieties of finger millet and the carbohydrate content ranges from 81.3 to 89.4%. Pore and Magar (1979) have summarized that carbohydrate, starch, reducing sugar contents of A16 and B11 finger millet varieties were 79.5 and 77.0%; 70.2 and 69.0% and 1.8 and 1.2%, respectively. They also reported that the hemicelluloses, cellulose and other carbohydrates were 7.5 and 6.8% respectively.

Table 3.1. Major Nutrients (%) of Finger Millet Varieties*

Sl.No.	Variety	Carbohydrate	Lipid	Protein	Reference
1	H22	84.5	1.5	7.7	Subramanyan et al, 1955
2	B11	87.5	1.9	8.2	Pore and Magar 1977.
3	A16	88.9	2.2	6.8	
4	E31	88.4	2.3	7.1	
5	White Nagali	88.4	2.4	6.9	
6	Red compact panicle variety	87.3	2.4	7.7	
7	Red loose panicle variety	89.4	2.4	5.8	
8	Apk 2	83.5	1.8	11	
9	Apk7	85.5	2.6	9.1	
10	PR 202	87.5	2.1	7.5	
11	PR248	86.5	2.4	8.3	
12	Vzm1	81.3	3	12.8	
13	Vzm2	82.8	2.1	11.8	
14	VR7	85.5	2	9.4	
15	VR18	84.2	1.6	11	
16	C157	84	2.4	11.2	
17	CR652	88.1	2	7	
18	CR312	87.7	2	7.2	
19	CO7	88.2	1.9	6.6	
20	Hamsa	85.7	1.6	10.4	
21	Purna	86.9	1.4	9.9	
22	HP7	85.1	1.6	10.9	
23	HP18	84.7	1.6	11.2	
24	HP35	85.3	1.7	10.7	
25	HP62	86.1	2	8.9	
26	HP69	85.2	2.1	10	
27	HP71	85.3	2	10	
28	HP81	84.4	2	10.8	
29	HP83	84.9	1.9	10.5	
30	HP86	84.9	1.9	10.5	
31	HP90	85.7	1.6	10.1	
32	HP91	84.9	1.3	11.1	
33	HP94	83.8	2.4	11.3	
34	HP102	83.2	2.7	11.6	
35	HP103	84.3	1.6	11.4	
36	HP105	85.2	1.3	11	
37	HP114	85.9	1.6	9.8	

Table 3.1. (Continued)

Sl.No.	Variety	Carbohydrate	Lipid	Protein	Reference
38	Chikwelekwele (Brown)	_	1.1	7.3	Mwabene, 1989.
39	Makukulu (B)	_	1.1	7.4	
40	Amakazi (B)	_	1.1	7.4	
41	Maulutila (W)	_	1.1	8.5	
42	Tukuyu (Black)	_	1.1	8.0	
43	CO10	81.8	1.5	9.5	Ravindran,1991
44	KM1	81.1	1.6	10.6	
45	MI302	81.8	1.6	9.2	
46	Ethiopia wild (KH2396)	71.5	4.2	11.0	Barbeau et al, 1993
47	Kenya local (KH265)	75.3	4.5	7.5	
48	Ethiopia local (KH2301)	74.0	4.3	8.7	
49	Madras local(KH2388)	74.2	4.4	9	
50	Bengal local(KH2324)	72.8	4.2	11.1	
51	Kashmir local(KH2339)	74.2	3.9	9.7	
52	Bihar local(KH2333)	72.4	3.8	11.7	
53	INDAF5	_	_	8.7	Udayasekhar Rao, 1994
54	WR9	_	_	12.3	
55	PES400	_	1.3	7.0	Sangita and Sarita, 2000
56	PES176	_	1.0	8.9	
57	PES110	_	1.4	7.4	
58	PRES4	_	1.4	7.4	
59	PRES5	_	2.0	8.4	
60	Mbeke	82.7	3.12	8.91	Shayo et al, 2001
61	Bambare	80.1	3.2	12.5	
62	Muhone	81.4	3.05	10.7	
63	Ukusi	81.1	3.31	10.9	
64	Ulyo	81.5	3.17	10.8	
65	Lanet (Kenyan variety)	74.7	1.6	6.5	Mbithi et al, 2000
66	VL46	80.1	1.59	8.5	Premavalli et al, 2006
67	VL49	75.1	1.04	7.7	
68	VL204	69.3	1.43	7.48	
69	Indaf5	70.2	1.31	7.5	
70	Indaf8	75.3	1.29	6.02	
71	Indaf9	74.2	1.79	5.64	
72	Indaf11	67.9	1.53	7.90	
73	GPU28	75.6	1.91	8.78	
74	MR1	77.8	1.70	5.23	
75	Indaf15	76.51	1.08	7.52	Desai et al,

Sl.No.	Variety	Carbohydrate	Lipid	Protein	Reference
					2010
76	GPU28	61.0	1.5	7	Usha and Malleshi, 2011

* Compiled from various sources.

Table 3.2. Reducing, Non-reducing and Total Sugar contents of Finger Millet Varieties*

Sl. No.	Varieties	Reducing Sugar	Non-Reducing Sugar	Total Sugar	Reference
1	Madras, local	0.43	-	1.4	Usha et al, 1996
2	Vl146	1.10	2.48	3.58	Mbithi et al, 2000 Premavalli et al,2006
3	VL149	1.44	2.68	4.12	
4	VL204	1.47	2.76	4.24	
5	Indaf5	1.10	2.37	3.47	
6	Indaf8	0.86	3.47	4.34	
7	Indaf9	1.41	2.67	4.09	
8	Indaf11	1.11	2.02	3.13	
9	GPU28	1.12	2.09	3.22	
10	MRI	0.86	2.51	3.37	
11	HR911	1.54	2.91	4.46	

* Compiled from various sources.

Finger millet has total starch 60 to 65%, Pentosan 6.2 to 7.2%, Cellulose 1.4 to 1.8%, lignin 0.04 to0.06% and free sugars 0.59 to 0.69% (Moruzzi, 1991). Premavalli et al (2004) reported nutrient profile in three hilly varieties and seven base varieties with starch content ranging between 67.9 to 80.5% with amylose 14.2 to 17.9% and amylopectin 82.1 to 86.1%. It was observed that the hilly and base varieties did not have influence on these fractions. Some of the researchers have studied the sugar fractions of finger millet. Antony et al (1996) have reported 1.4% total soluble sugar and 0.43% reducing sugar in a Madras local variety of fingermillet. The free sugars (mg/g dry matter) present were D-xylose (1.12), D-fructose (0.09), D-glucose (0.09) and sucrose 0.12). Subbarao et al (2004) reported sugar composition of cold water soluble polysaccharides of INDAF-15 cultivar which contained rhamnose (8.5%), arabinose (26.7%), xylose (10.6%), mannose (5.5%), galactose (17.9%) and glucose (30.7%). Premavalli et al (2006) reported that the total sugars ranged from 3.13 to 4.46% with the reducing sugar in the range of 0.86-1.54% among the nine varieties. The altitude had no effect on sugar levels (Table 3.2). The major sugars identified in a brown variety GPU-28 (Usha and Malleshi 2011) were sucrose, fructose and maltose amounting to 1.0.85%. The free sugar contents of 70% aqueous ethanol extract of native GPU-28 flour were glucose0.387, fructose0.105, maltose0.157, sucrose0.225, ribose0.086, others0.125g/100g. Further the cold water soluble non starch polysaccharides contained ribose, arabinose, xylose, mannose and maltose. The above studies reveal the compositional differences in the carbohydrate components of finger millet. Though the sugar components show variation with the varieties, it is present in the grain at lower levels and the impact may be on non-enzymatic browning reactions during processing. However, these reactions may not influence much on the product sensory quality in case of finger millet products because of its native dark colour of grains. But these reactions are of importance in flavour generation specially during high temperature processing. Hence, the starch component is relatively more important and the starch fractions, their digestibility have an influence on the metabolic reactions. The carbohydrate components viz. sugars, starch, dietary fibre in the fingermillet recorded 5-6% difference amongst varieties and the altitude had no much influence on these parameters (Premavalli et al, 2004). Usha and Malleshi (2011) have reported 61% carbohydrate with 16.7% amylose in GPU28 variety. The detailed discussion on starch fractions and dietary fibre, their nutritional importance and effect of processing does not fall in purview of this chapter and it will be discussed in specific chapters.

2. Proteins

Protein being an important nutrient in the diet, any source of protein adds to the strength of the diet. But the quality and quantity of protein makes the difference for protein food. Finger millet has lower protein content and is known to be incomplete protein because of deficiency in lysine and threonine which are essential amino acids. However, it is a good source of glutamic acid. The prolamine called eleusinin is superior to gliadin of wheat as regards cystine, tyrosine and tryptophan (Wealth of India, 1952). The protein content of 76 varieties of finger millet is given in Table 3.1. The protein content ranges from 5.23 to 12.8% which clearly reflects the varietal difference. Rachie and Peters (1977) reviewed that fingermillets contain around 7% proteins while, Pore and Magar (1977) reported a range of 5.8 to 12.87 protein in 36 varieties of finger millet. Rao and Mushenga (1985) have discussed the fingermillet crop of Zimbabwe as compared to Indian cultivars and also reported wide variation of protein from 5.8 to 14.2% in fingermillet. Shukla at al (1986) in their study on 8 high yielding varieties reported the range of protein from 3.7 to 4.6%. Barbeau and Hilu (1993) reported protein in two wild and 8 domesticated cultivars of fingermillet wherein the mean values ranged from 7.5 to 11.7%. Udayasekhara Rao (1994) reported white fingermillet had higher protein of 12.3% than brown fingermillet with 8.7% protein. National Research Council (1996) reported that nutritional promise of fingermillet in terms of its various nutrients. The fingermillet on an average contained 323-350 Kcal food energy for 100g in which protein content ranged from 6 to14%. Sankara et al (1998) reported that among the thirty six genotypes of fingermillet, white seeded genotypes showed higher protein contents, while brown seeded genotypes exhibited wide range of variation. Malleshi and Kloptenstein (1998) reported 7% protein in Indaf variety. Then, Premavalli et al (2006) have compared the hilly and base varieties and reported 7.48 to 8.5% protein in hilly varieties and 5.23 to 8.78% in base varieties.

Some of the researchers have studied mainly on the protein levels of fingermillet varieties. The studies earlier to 1975 have been given briefly in Table 3.3 and the later period research result of 40 varieties has been listed in Table 3.4. In the earlier period as reported in 136 varieties of fingermillets from 1954, protein content ranged from 4.5 to 12.7%. Thirty African varieties was studied by Shepherd et al (1971) declare 7.5 to 9.5% proteins while in Phillippines two varieties (Solpico and Yambao, 1966) the protein content is 9.1 and 11.9%. In recent years the protein content ranges from 6.36 to 12.3% over the decades, though new hybrid varieties are cultivated, the impact is not much on proteins levels. Thus the finger millet varietal difference shows the protein levels of 4.5 to 12.8% which reflects clearly the higher protein varieties too. The study of various fractions, amino acid patterns leads to better understanding of quality of proteins. Indira and Naik (1971) have reported 21.5% prolamine, 17.9% globulin, 12.7% albumin and 12.2% glutelin in fingermillet and discussed about the amino acid composition of these fractions. Virupaksha et al (1975) have reported that white varieties of finger millet have more prolamines but less glutalin than brown varieties. Geeta et al (1978) reported that among the five cultivars of finger millet prolamine fraction was highest followed by glutelin fraction and albumin-globulin fraction. They have resolved protein subunits of finger millet varieties according to their molecular weight using sodium dodecyl sulphate-polyacrylamide gel electrophoresis (SDS-PAGE). They reported that varietal differences exist in the protein compositions of finger millet varieties. The molecular weight distribution of the protein subunits in the albumin-globulin, prolamine and glutelin fractions show many differences between parental and cross breed varieties. These differences were greater in the albumin-globulin and glutenin fractions than in the prolamine fractions. Venkanna Babu et al (1987) have reported protein fractions in six varieties of fingermillet. Albumins ranged from 0.7 to 1%, globulins from 1.5 to 1.7%, glutelins from 4.3 to 6.6% with prolamines from 1.9 to 3.4%. Premavalli et al (2006) have reported on protein fractions of 11 varieties of finger millet. Amongst the varieties, water, alcohol and salt soluble proteins ranged from 1.4 to 4.42%, 1.17 to 3.09% and 1.26 to 3.61% respectively (Table 3.4). Prolamines, alcohol soluble fraction pattern was found to be not maximum as per earlier reports but is in correspondence with Venkanna Babu et al (1987) reflecting that prolamine is one of the major fractions of fingermillet. The altitudes did not influence the protein fractions. The amino acid profile provides a more clear picture on the quality of protein. Several researchers have reported on amino acid profile of fingermillet varieties and expressed in different units. After conversion into common unit, the amino acid profile of 7 varieties is given in Table 3.5. In all the varieties most of the essential amino acids are present. The amino acids in common reported are isoleucine, leucine, lysine, methionine, phenyl alanine, threonine valine which are essential amino acids. But are found in different proportions where leucine is highest in concentration followed by phenyl alanine. Lysine is said to be present in lower quantities in finger millet, and in a similar manner, though these samples are less in lysine but not the least. Among other amino acids, fingermillet is a good source of glutamic acid which is present to the tune of 16 to 21% as compared to 1 to 7.5% of all other amino acids.

Table 3.3. Protein Content in Indian Finger Millet Varieties*

Sl.No.	No. of Varieties	Protein, %	References
1.	8	8.2 – 12.2	Kadkol and Swaminathan, 1954
2.	22	9.9 – 12.1	Venkataramana and Krishna Rao, 1961
3.	30	7.7 – 11.4	Sree Ramulu and Mariakulandai, 1964
4.	10	7.9 – 11.6	Mahudeswaran et al, 1966
5.	25	4.5 – 7.6	Krishnamurthy, 1968
6.	19	8.7 – 12.7	Kempanna and Kavalappa, 1968
7.	5	6.2 – 9.5	Indira and Naik, 1971
8.	2	7.3 and 10.8	Kamalanathan et al, 1971
9.	15	4.7 – 9.9	Balakrishna Rao et al, 1973

* Compiled from various sources.

Table 3.4. Protein Profile of Finger Millet Varieties*

Sl.No.	Variety	Protein,%	Reference
1.	VR 2506	11.1	Venkannababu et al, 1987
2.	VZM 1	9.8	
3.	VZM 2	10.7	
4.	C 157	12.1	
5.	APK 2	10.5	
6.	Godavari	8.0	
7.	Indaf5	8.70	Udayasekhara Rao, 1994
8.	WR9	12.30	
9.	IE2300	9.70	Sonnad, 2005
10.	IE2303	7.59	
11.	IE2835	9.97	
12.	IE2837	8.45	
13.	IE2860	8.13	
14.	IE2861	9.53	
15.	IE2863	10.52	
16.	IE2872	11.03	
17.	IE2875	9.22	
18.	IE2887	7.93	
19.	IE2889	8.08	
20.	IE2903	11.40	
21.	IE2906	7.52	
22.	IE2907	8.30	
23.	IE2976	8.49	
24.	IE2995	11.22	
25.	WRC 112	8.02	
26.	OUAT2	9.11	
27.	Dapoli white	9.64	
28.	Indaf11	8.70	
29.	GPU28	8.82	
30.	GPU26	8.93	
31.	Nigerian Black variety	8.47	Glew et al, 2008
32.	Boneya	10.50	Shimelis et al, 2009
33.	Padet1	9.86	
34.	PBL1	6.56	
35.	PBL2	6.71	
36.	PBL3	6.64	

Table 3.4. (Continued)

Sl.No.	Variety	Protein,%	Reference
37.	PBR1	6.70	
38.	PBR2	6.80	
39.	PBR3	6.36	
40.	Tedesse	7.61	

* Compiled from various sources.

Table 3.5. Protein Fractions (%) In The Varieties of Native Finger Millet*

Sl.No.	Varieties	Salt Soluble	Water Soluble	70% Alcohol Soluble
1.	VL146	1.92	4.42	2.05
2.	VL149	1.26	1.89	1.81
3.	VL204	2.04	3.04	2.30
4.	Indaf5	2.98	1.78	2.10
5.	Indaf8	2.44	2.04	1.62
6.	Indaf9	2.49	1.40	2.08
7.	Indaf11	3.61	2.87	1.17
8.	MR1	2.38	1.63	1.18
9.	GPU28	3.61	1.60	3.09
10.	HR-911	2.78	2.21	2.96
11	Local	2.75	1.66	2.10

* Premavalli et al, 2006.

Virupaksha et al (1975) reported amino acid composition of whole seed protein of 12 fingermillet varieties (Table 3.6) and the results showed wide variations.

Amongst varieties glutamic acid, proline, valine, isoleucine and leucine exhibited a wider range of values compared to the other amino acids. They found that there was gross similarity between the amino acid spectrum of endosperm meal proteins and whole grain protein, among the four varieties Hamsa, Purna, HPW834 and HPB205. Therefore, the protein present in the endosperm meal is representative of the protein of the whole seed flour. Geeta et al (1978) have reported that amino acid composition of all the five varieties studied was generally similar but with a few notable exceptions in concentration of amino acids. Prolamines fraction has low level of lysine and very high levels of glutamic acid with high levels of proline, valine, alanine, leucine, and phenylalanine. Methionine level was higher in the prolamine fraction as compared to other fractions. Malleshi and Klopfenstein (1998) have reported the amino acid pattern of Indaf variety which shows the amino acids ranged from 3 to 8% with highest of glutamic acid at 21% level. But Thatam et al (1996) have found glutamic acid, alanine and proline as major amino aicds in one of the varieties. Prolamine being an important protein fraction of finger millet because of high biological value, Thatam et al (1996) have isolated and characterized the prolamines of finger millet and the comparative studies indicated that the major prolamines are similar to those of prolamines of maize, sorghum and teff. Mbithi et al (2002) have reported essential amino acids of Kenyan variety fingermillet which showed valine as highest followed by leucine, phenyl alanine, and others with lowest level of lysine. Glew et al (2008) have reported the amino acid pattern of black fingermillet and is a good source of essential amino acids except lysine.

Table 3.6. Amino Acid Profile of Finger Millet Varieties

Variety → / Amino acids ↓	Finger millet[a]	WR9[b]	INDAF5[b]	A16[c]	B11[c]	INDAF[d]	Ragi[e]
Cystine	2.6	-	-	-	-	-	2.24
Isoleucine	4.4	2.36	1.45	6.1	6.76	8.75	6.4
Leucine	9.5	5.28	5.28	9.5	10.5	2.74	11.04
Lysine	2.90	1.49	1.62	3.25	3.44	3.11	3.52
Methionine	3.10	2.49	2.64	3.6	3.0	3.39	3.36
Phenyl alanine	5.2	3.22	3.15	5.2	4.6	5.48	4..96
Threonine	4.2	2.72	3.73	3.1	3.2	4.27	3.84
Tryptophan	3.05	-	-	-	-	-	1.6
Tysrosine	-	2.60	2.30	5.0	5.3	3.86	3.52
Valine	6.61	2.50	2.32	5.77	6.86	4.84	7.68
Arginine	-	2.19	2.49	5.5	6.8	4.40	4.8
Aspartic acid	-	2.63	6.63	-	-	7.59	-
Glutamic acid	-	16.78	16.33	-	-	20.98	-
Glycine	-	2.17	2.56	-	-	4.57	-
Proline	-	5.57	4.99	-	-	6.37	-
Histidine	-	1.70	1.72	1.5	1.4	3.10	2.08
Serine	-	3.83	5.18	-	-	5.48	-
Alanine	-	4.10	4.61	-	-	-	-

[a]: FAO 1970, Indira and Naik, 1971.

[b]: Pore and Magar, 1979.

[c]:Udayasekhara Rao, 1994.

[d]:Malleshi and Kloptenstein, 1998.

[e]: Gopalan et al, 2004.

They concluded that black variety and more common tan coloured variety both contain only half the proportion of lysine recommended by WHO.

3. Lipids

Lipids though present in smaller quantities in cereals, the type of lipids and their fatty acid composition influences the nutritional strength on one hand and stability on the other hand. Finger millet as a grain has the stability for 30 years, but when ground into flour stability is only for 3 months and the lipid composition plays a role. The lipid content of 76 varieties has been compiled in Table 3.1 and it ranges from 1 to 4.5%. Kadkol and Swaminathan (1954) have reported 1.4 to 1.9% in 8 varieties. Indira and Naik (1971) have found 1.4 to 2.5% lipids in 5 varieties. In African varieties, the lipid content reported (Shepherd et al 1971) is 0.7 to 1.7% in 30 varieties study, where the lipid content seems to be low in some varieties. But in Phillippines' two varieties Salpico and Yambao (1966) have found 4.5 and 4.6% lipids. Kempanna and Kavallappa (1968) have studied varieties and found 1.3 to 1.7 lipid content in finger millet. Mahadevappa and Raina (1978) reported 1.85 to 2.1% total lipids in finger millet varieties. While Rachie and Peters (1977) have reported 1.5% lipids in finger millet. Shukla et al (1986) in their study on 8 high yielding varieties have reported fat content ranging from 1.2 to 2.3%. Sridhar and Lakshminarayana (1994) reported 2.2% free lipids in finger millet. Antony et al (1996) have reported 2.1% fat in the variety studied. Premavalli et al (2006) have reported in 9 varieties of finger millet. The free lipids in hilly varieties ranged from 1.4 to 1.5% while in base varieties it ranged from 1.3 to 1.9%. The lipid content in hilly varieties has been reported for the first time. The lipid content of hilly varieties was lower by 0.21%, which may be attributed to environmental/agronomic conditions, however, the change were statistically non-significant. Free fatty acid content is generally used to assess the quality of cereal grains (Premavalli et al, 1973) and raw finger millet, free fatty acids content ranged from 0.18 to 0.28%, while in hilly samples, it was 0.15 to 0.57% in base varieties.

The major lipid classes were separated by column chromatography and quantified gravimetrically. Mahadevappa and Raina (1978) reported 70 to75%, 10 to 12%, 5 to6% neutral, glyco and phospho lipids respectively in finger millets, while Sridhar and Lakshminarayana (1994) reported 80 to 83% neutral lipids, 6 to14% glycolipids and 5 to14% phospholipids, in three different types of millets. In Indaf 1 and Hamsa varieties, Mahadevappa and Raina (1984) have further separated the glyco and phospholipids by chromatographic methods and found monogalactosyldiglyceride and digalactosyl diglyceride as major glycolipids while, phosphotidyl choline and phosphotidyl ethanolamine as major phospholipids. Wadikar et al (2007) have reported neutral lipids 74 to 77% with 17 to 21% glycolipids and 2 to 4.8% phospholipids, in hilly varieties, while in base varieties, it ranged from 70 to 85%, 12 to 28% and 0.5 to 5.36% respectively, indicating the variation in lipid fractions amongst varieties. However, the variation was statistically non-significant. The phospholipids were slightly lower in most of the varieties, while on the whole the lipid classes followed similar pattern.

Fatty acid profile gives the information about the essential fatty acids present and their stability characteristics in processed foods. The fatty acid composition of raw finger millet varieties have been studied little more widely compared to lipid classes. The fatty acid composition of finger millet varieties is given in Table 3.7.

Mahadevappa and Raina (1978) have studied seven varieties and reported only four fatty acids i.e. palmitic, oleic, linoleic and linolenic acids, out of which three are major acids and linolenic acid is present to the tune of 1.1 to 1.8%. Mahadevappa and Raina (1984) have reported seven fatty acids with palmitic, oleic and linoleic acids as major components.

Wadikar et al (2007) have reported fatty acid composition of 9 Indian varieties of finger millet out of which three i.e. VL146, VL149, VL204 were hilly varieties, while the rest were base varieties. Commonly, palmitic, stearic, oleic, linoleic and linolenic acids were found in all varieties but caproic and lauric acids though present in very small quantities were dependent on the varieties. Glew et al (2008) have reported six fatty acids in Nigerian varieties and palmitic, stearic, oleic, linoleic and linolenic were the main acids. Uhsa and Malleshi (2011) have reported four fatty acids in GPU28 variety and the same acids were reported by Wadikar et al (2007) in GPU28. But, the concentration of acids varied by 2 to 3% except for linoleic acid. Thus the fatty acid profile of finger millet revealed that finger millet contains 20.25 to 27.38% palmitic acid, traces to 2.13% stearic acid, 43.89 to 51.5% oleic acid, 20.15 to 26.02% linoleic acid and 1 to 4.25% linolenic acid reflecting 21 to 30% essential fatty acids and the compositional changes were variety dependent. Wadikar et al (2007) study on hilly varieties is the first report and relatively linolenic acid is slightly higher in these varieties.

Table 3.7. Amino Acid Composition of the Finger Millet Whole Seed Protein*

Variety → / Amino acids ↓	EC 4840	HPB 76	HPB 205	HPB 18	Purna	HPB 23-6	ECW 854	HPW 274	ECW 955	Hamsa	HX 799	HPW 834
Lysine	3.72	3.39	4.53	4.78	3.47	5.52	2.89	2.95	3.87	3.81	3.79	4.56
Histidine	3.03	2.57	2.97	3.17	2.35	3.98	2.44	1.65	2.98	2.70	3.14	2.445
Arginine	6.08	4.78	5.78	5.28	4.27	6.90	4.06	3.84	5.60	5.16	5.02	6.13
Aspartic acid	8.20	6.46	7.50	7.55	8.77	7.75	7.53	6.27	10.02	7.21	6.98	8.71
Threonine	5.72	4.11	4.83	5.05	5.77	5.49	3.92	4.38	5.58	4.99	5.11	5.64
Serine	7.20	6.14	6.72	7.10	7.86	7.05	6.20	5.52	8.71	6.89	5.99	8.66
Glutamic acid	36.47	26.49	25.89	23.13	28.81	27.13	34.30	35.90	37.82	32.38	27.14	31.95
Proline	9.68	4.20	6.65	5.29	8.02	5.75	5.26	5.88	4.34	6.16	4.89	6.99
Glycine	4.72	4.00	4.83	4.71	5.27	4.89	3.73	3.92	4.70	4.79	5.01	5.75
Alanine	8.23	7.26	7.71	6.84	8.78	8.04	7.46	6.28	8.73	8.52	8.19	8.99
Valine	9.69	7.48	7.39	6.80	8.34	7.41	7.47	10.40	9.53	7.88	6.98	8.26
Methionine	3.02	2.32	3.38	3.02	2.24	3.29	1.81	2.71	3.63	2.02	3.36	4.26
Isoleucine	5.42	4.23	4.70	4.49	5.30	5.07	4.17	3.83	6.45	4.51	4.46	4.31
Leucine	14.68	12.03	12.63	10.85	15.13	12.85	13.14	13.96	16.17	14.13	12.33	14.89
Tyrosine	4.56	3.77	3.85	3.22	3.90	3.93	3.07	3.08	5.48	3.01	3.28	3.29
Phenylalanine	7.91	6.28	6.35	5.94	7.00	7.03	5.35	5.13	8.43	5.90	5.85	5.37

* Grams per 16 g of nitrogen, Virupaksha et al, 1975.

Table 3.8. Fatty Acid Composition (%) of Raw Finger Millet Varieties*

Sl No.	Varieties	C_{12}	C_{14}	C_{16}	C_{18}	$C_{18:1}$	$C_{18:2}$	$C_{18:3}$	Refernce
1	PR202	_	_	25	traces	49.8	24.1	1.1	Mahadevappa and Raina, 1978
2	HPB76	_	_	25.6	traces	49.0	24.3	1.1	
3	ECW1360	_	_	25.0	traces	50.0	24.0	1.0	
4	EC 4840	_	_	26.0	traces	48.5	24.5	1.0	
5	INDAF1	_	_	25.0	traces	50.0	23.2	1.8	
6	HAMSA	_	_	23.0	traces	51.5	24.0	1.5	
7	HES 929	_	_	23.5	traces	50.0	25.0	1.5	
8	Madras local	0.20	-	5.29	8.99	_	5.98	0.89	Antony et al, 1996
9	VL146	-	0.15	27.22	0.76	47.48	20.15	4.11	Wadikar et al, 2007
10	VL149	0.15	-	24.37	0.92	43.89	26.02	4.25	
11	VL204	0.24	-	27.38	0.56	44.58	22.54	3.73	
12	Indaf 5	0.05	0.06	23.93	1.58	46.15	25.19	2.81	
13	Indaf 8	-	0.62	25.99	1.99	47.28	20.81	3.18	
14	Indaf 9	0.07	0.24	23.81	1.21	47.56	23.42	3.55	
15	Indaf 11	0.2	0.16	26.52	1.56	46.32	22.53	2.50	
16	MR1	0.6	0.11	25.43	1.37	48.17	22.28	2.5	
17	GPU28	-	-	23.24	1.42	49.55	21.68	4.0	
18	Nigerian black var.	0.20	_	22.7	2.13	49.13	20.2	3.74	Glew et al, 2008
19	GPU28	_	_	20.25	-	51.13	24.42	4.07	Usha and Malleshi, 2011

* Compiled from various sources.

Table 3.9. Mineral Composition of Finger Millet Varieties[#]

Sl.No.	Variety	Minerals, mg %									References
		K	Na	Mg	Ca	P	Mn	Zn	Cu	Fe	
1.	B11	495*	10*	131	284	131	0.6	-	0.33	3.4	Pore and Magar, 1977 *Pore and Magar, 1979
2.	A16	365*	20*	105	320	116	0.1	-	0.28	6.7	
3.	E31	-	-	110	292	125	-	-	-	6.9	
4.	White Nagali	-	-	110	303	219	2.7	-	0.66	5.4	
5.	Red compact panicle variety	-	-	120	320	212	2.8	-	0.65	8.7	
6.	Red loose panicle variety	-	-	136	378	179	3.1	-	1.09	8.8	
7.	Akp 2	-	-	251	310	309	1.9	-	1.55	22.8	
8.	Akp7	-	-	233	378	291	1.4	-	1.11	8.9	
9.	PR 202	-	-	221	460	283	3.2	-	0.93	15.4	
10	PR248	-	-	229	520	232	3.6	-	1.12	18.4	
11.	Vzm1	-	-	221	348	277	1.5	-	0.87	9.4	
12.	Vzm2	-	-	209	330	279	1.5	-	0.86	19-	
13.	VR7	-	-	251	480	404	3.1	-	0.48	13.2	
14.	VR18	-	-	221	290	237	1.8	-	1.1	8.8	
15.	C157	-	-	204	432	238	1.9	-	1.32	5.4	
16.	CR652	-	-	220	410	250	-	-	-	6.8	
17.	CR 312	-	-	206	390	220	-	-	-	8.9	
18.	CO7	-	-	209	422	268	-	-	-	7.4	
19.	Hamsa	-	-	169	345	261	1.3	-	1.3	7.3	
20.	Purna	-	-	164	261	273	1.1	-	1.1	5	
21.	HP7	-	-	144	284	288	1.2	-	0.8	4.6	

Sl.No.	Variety	Minerals, mg %									
		K	Na	Mg	Ca	P	Mn	Zn	Cu	Fe	References
22.	HP18	-	-	154	294	306	1.7	-	1.1	4	
23.	HP35	-	-	156	275	276	2	-	1	6.7	
24.	HP62	-	-	157	331	270	0.9	-	0.7	4.8	
25.	HP69	-	-	170	279	306	1.2	-	1.2	4.3	
26.	HP71	-	-	106	333	256	1.4	-	0.5	4.5	
27.	HP81	-	-	167	320	278	1.8	-	0.7	4.9	
28.	HP83	-	-	158	289	280	1.4	-	1	5.2	
29.	HP86	-	-	165	377	314	1.8	-	1.1	5.6	
30.	HP90	-	-	176	318	303	1.2	-	1.2	4.8	
31.	HP91	-	-	160	283	251	1.3	-	1.2	4.9	
32.	HP94	-	-	152	287	254	1.2	-	1	4.6	
33.	HP102	-	-	165	320	282	1.3	-	1.2	5.2	
34.	HP103	-	-	176	324	278	1.4	-	1.1	7.2	
35.	HP105	-	-	152	321	249	1.3	-	1	4.8	
36.	HP114	-	-	162	259	264	2.1	-	1.2	6.3	
37.	JNR852	197.33	12.00	176.96	505.60	310.30	37.74	4.70	0.95	10.88	Shukla et al, 1985
38.	JNR981	194.66	09.30	170.63	526.66	319.20	44.77	4.34	0.75	10.24	
39.	JNR1008	264.00	13.30	198.92	542.46	316.60	24.79	4.80	0.10	12.16	
40.	PR202	173.33	08.80	186.82	489.00	331.86	56.61	2.67	0.68	09.60	
41.	RAU6	194.66	11.00	180.12	428.00	267.66	43.66	3.39	0.90	12.16	
42.	RAU8	232.00	11.00	186.82	428.00	267.66	55.87	4.00	0.85	09.02	

Table 3.9. (Continued)

Sl.No.	Variety	Minerals, mg %									
		K	Na	Mg	Ca	P	Mn	Zn	Cu	Fe	References
43.	HR374	162.66	13.28	176.96	383.00	276.92	30.74	3.50	0.90	10.33	
44.	PES176	208.00	08.76	195.20	404.00	258.96	39.00	3.12	0.80	10.33	
45.	KM13	168.00	09.11	183.02	409.00	248.71	45.89	4.53	0.68	11.98	
46.	Humsa	184.78	12.00	167.82	450.00	230.76	45.00	4.01	0.88	10.66	
47.	WR13	189.03	11.91	176.00	389.73	357.20	43.60	3.09	0.82	12.08	
48.	CO9	173.33	11.00	168.28	331.80	349.60	52.00	4.56	0.85	05.76	
49.	WR5	196.28	12.00	162.28	421.30	298.96	63.27	4.48	0.72	11.64	
50.	WR4	190.29	08.93	179.00	395.00	357.20	46.99	3.04	0.56	7.68	
51.	VR 2506	-	-	-	359.5	25	-	-	50.0	14.8	Venkannababu et al, 1987
52.	VZM 1	-	-	-	390.6	30.8	-	-	61.6	11.4	
53.	VZM 2	-	-	-	332.1	26.4	-	-	47.5	7.7	
54.	C 157	-	-	-	336.1	29.3	-	-	55	3.3	
55.	APK 2	-	-	-	327	28.0	-	-	55	10.9	
56.	Godavari	-	-	-	293.9	29.5	-	-	59.1	6.1	
59.	Chikwelekwele (Brown)	-	-	-	350	250	-			25	Mwabene,1989
60.	Makukulu (B)	-	-	-	310	026	-	-	-	28	
61.	Amakazi (B)	-	-	-	320	270	-	-	-	25	
62.	Maulutila (W)	-	-	-	360	300	-	-	-	28	
63.	Tukuyu (Black)	-	-	-	450	330	-	-	-	37	
64.	CO10	560	90	140	250	280	6	2	3	5	Ravindran,1991
65.	KM1	580	80	130	250	200	4	2	4	5	

Sl.No.	Variety	Minerals, mg %									References
		K	Na	Mg	Ca	P	Mn	Zn	Cu	Fe	
66.	MI302	570	50	130	240	230	4	2	4	4	
67.	Indaf5	-	-	-	-	202	-	2.1	-	4.4	Udayasekhara Rao,1994
68.	WR9	-	-	-	-	197	-	2.4	-	12.0	
69.	Madras Local	-	-	-	250.9	127.8	4.54	1.11	0.26	1.71	Antony and Chandra, 1998
70.	PES400	-	-	-	503.7	-	-	2.6	-	14.8	Sangita Kumari and Sarita, 2000
71.	PES176	-	-	-	549.0	-	-	4.3	-	17.3	
72.	PES10	-	-	-	487.0	-	-	2.4	-	15.8	
73.	PRES4	-	-	-	547.7	-	-	3.0	-	12.1	
74.	PRES5	-	-	-	532.3	-	-	4.3	-	10.2	
75.	Fingermillet	408	11.0	137.0	344	283	-	2.3	-	3.9	Gopalan et al, 2004
76.	Mbeke	-	-	-	330	240	-	-	-	-	Shayo et al, 2001
77.	Bambare	-	-	-	100	350	-	-	-	-	
78.	Muhone	-	-	-	170	450	-	-	-	-	
79.	Ukusi	-	-	-	340	120	-	-	-	-	
80.	Ulyo	-	-	-	30	350	-	-	-	-	
81.	VL146	-	-	-	350	-	-	-	-	17.1	Majumdar et al, 2006
82.	VL149	-	-	-	358	-	-	-	-	17.2	
83.	VL204	-	-	-	366	-	-	-	-	18.5	
84.	Indaf5	-	-	-	378	-		-	-	13.0	
85.	Indaf8	-	-	-	372	-	-	-	-	13.1	

Table 3.9. (Continued)

Sl.No.	Variety	Minerals, mg %									References
		K	Na	Mg	Ca	P	Mn	Zn	Cu	Fe	
86.	Indaf9	-	-	-	375	-	-	-	-	13.6	
87.	Indaf11	-	-	-	412	-	-	-	-	11.3	
88.	GPU28	-	-	-	375	-	-	-	-	14.3	
89.	MR1	-	-	-	370	-	-	-	-	13.2	
90.	HR911	-	-	-	380	-	-	-	-	15.8	
91.	Local	-	-	-	350	-	-	-	-	16.8	
92.	INDAF5	1070	0.90	110.0	293	234	140	2.31	0.5	4.44	Sashi et al, 2007
93.	GPU28	1160	0.95	130.0	298	250	29.2	2.26	0.75	6.35	
94.	ML197	328	0.70	90.0	359	248	-	1.83	-	4.43	
95.	ML365	328	0.70	80.0	365	259	-	1.76	-	5.52	
96.	ML53	350	0.95	96.0	346	240	-	2.14	-	7.31	
97.	ML426	360	0.85	99.0	327	285	-	2.14	-	4.00	
98.	ML31	334	0.90	84.65	264	292	-	1.96	-	3.60	
99.	ML322	294	0.60	66.0	281	280	-	1.87	-	3.84	
100.	Nigerian Black variety	541	14.8	174	401	276	-	2.71		18.2	Glew et al, 2008
101.	Boneya	-	-	78.0	50.66	3.46	18.87	0.97	0.18	6.01	Shimelis et, al 2009
102.	Padet1	-	-	165	248.00	135	17.61	2.56	0.73	4.59	
103.	PBL1	-	-	201	319.0	141	48.43	1.91	0.66	34.15	
104.	PBL2	-	-	186.33	299	147	445.39	1.57	0.61	26.62	
105.	PBL3	-	-	193	307	133	48.08	1.69	0.60	27.31	
106.	PBR1	-	-	174	289	126	17.80	1.72	0.73	53.13	

Sl.No.	Variety	Minerals, mg %									
		K	Na	Mg	Ca	P	Mn	Zn	Cu	Fe	References
107.	PBR2	-	-	179.33	277	140	18.98	1.75	0.79	53.39	
108.	PBR3	-	-	179.33	271	125	19.86	1.71	0.55	30.66	
109.	Tedesse	-	-	163	248	146	20.17	2.31	0.72	6.25	
111.	INDAF15	-	-		359.4	284.3				13.7	Desai et al, 2010
112.	GPU28	-	-	-	321	244	-	2.10	1.6	6.0	Usha and Malleshi, 2011

Compiled from various sources.

4. MINERALS

Minerals are known for their physiological actions and are required in small quantities thereby they are called as minor nutrients. Finger millet is a good source of minerals in general, and calcium and iron in particular. Phosphorous and magnesium are present in lesser quantities than calcium. Several researchers have studied on the mineral composition of finger millet varieties. The total ash content ranges from 2.4 to 4%. Generally, the major elements calcium, phosphorous and iron have been studied, but in the later years other trace elements have also gained importance in analysis. The earlier work refers to 1920-30's, which has stated that Ragi is rich in calcium, phosphorous, iron and also contain sulphur and zinc. The iodine content of 101µg/Kg has been reported to be the highest among food grains (Wealth of India, 1952.). Kadkol and Swaminathan (1954) study on 6 brown and two white varieties where calcium ranges from 301 to 495 mg%, phosphorous 228 to 370 mg%, iron 5.9 to 6.9 mg% and relatively, the white variety called majjige ragi had highest calcium levels. Subramanyan et al (1955) reported 502 mg% calcium and 325 mg% phosphorous in the variety studied. Lower calcium level of 389 mg% with phosphorous 239% has been reported by Joseph et al (1958). Sree Ramulu and Maria Kulandai (1964) as per their analysis of 30 varieties for calcium and phosphorous revealed very high levels recording 613 to 1116 mg% calcium and phosphorous 639 – 1002 mg%. Krishnamurthy (1968) in his study on 25 varieties stated 289 to 445 mg% calcium. The work of Deosthale et al (1970) on 20 high yielding varieties reported 253 to 661 mg% calcium, 204 – 330 mg% phosphorous and 1.3 to 17.6 mg% iron. Phosphorous and magnesium levels were relatively very narrow. Copper, molybdenum and manganese ranged from 0.32 to 1 mg%, 0.011 to 019 mg% and 0.0011 to 0.021 mg% respectively. The white variety CO9 had 417 mg% calcium while brown variety had 344 mg%; but iron was higher in brown samples recording 17 mg% as compared to white ragi, 12 mg% as per Kamalanathan et al (1971). Indira Naik (1971) in 5 varieties found calcium levels to be 272 to 352 mg%, phosphorous 242 – 284 mg% with iron 11-18 mg%. Balakrishna Rao et al (1973) studied 15 varieties and reported calcium from 203 to 690 mg%, iron 2.5 to 19.9 mg% with phosphorous from 227 to 470 mg%. Kumaraswamy and Venkataramana (1974) studied only calcium levels in 10 varieties which ranged from 164 to 369 mg% which is relatively having lower calcium varieties as compared to other studies. In the subsequent years i.e. from 1977 to date, fingermillet minerals were studied exhaustively in most of the cultivated varieties. The list of 109 varieties along with minerals is given in Table 3.9. Pore and Magar (1979) have reported potassium and sodium in ragi for the first time, B11 variety had 495 mg% potassium while A16 variety had 365 mg% with low levels of sodium 10 and 20 mg% respectively. The same authors have analysed 36 varieties of ragi and reported calcium 259 to 520 mg%, 116 to 404 mg% phosphorous, 3.4 to 22.8 mg% iron, 105 to 251 mg% magnesium with copper and manganese in smaller quantities. Shukla et al (1985), in their study reported maximum minerals with zinc as an additional element present at 3.04 to 4.7 mg% levels. In these varieties potassium present at 173 to 264 mg% was relatively low as compared to earlier study. However, sodium levels of 8.76 to 13.3 mg% are at par with the other varieties reported. The calcium levels ranged from 331 to 542 mg% with phosphorous 248 to 357 mg% and iron 5.76 to 12.16 mg%. Copper was present at very low levels of 0.1 to 0.95 mg% with magnesium levels of 162.3 to 198.9 mg%. Venkannababu et al (1987) study on 6 varieties showed very high levels of copper ranging from 47.5 to 61.6 mg% with calcium, phosphorous and iron levels as given in Table 3.9. The study of Udayasekhara Rao and Deosthale (1988) was about malting characteristics of 3 white and 12 coloured varieties; however iron and zinc of native varieties have been reported on average of 4.95 mg% iron and 2.05 mg% of zinc. Tanzanian five varieties reported by Mwabene (1989) contained 310 to 450 mg% calcium, 250 to 330 mg% phosphorous and 25 to 37 mg% iron and the highest level of iron so far reported in finger millet. Ravindran (1991) have reported 9 elements in 3 varieties, but the levels of calcium and iron were lower with higher levels of sodium and potassium comparatively to other researchers. Barbeau and Hilu (1993) reported 2 wild and 8 domesticated cultivars of finger millet wherein the values of calcium ranged from 376 to 515 mg% and iron from 3.7 to 6.8 mg%. Udayasekhara Rao (1994) studied 2 varieties for phosphorous, zinc and iron along with antinutrient factors. Antony and Chandra (1998) reported 7 elements in local variety, but calcium and iron are lower levels in general. Sangita and Sarita (2000) have reported on calcium 487 to 548 mg%, iron 10.2 to 17.3 mg% with zinc at 2.3 to 4.3 mg% in 5 varieties and in this study, calcium levels are found to be high in these varieties. Gopalan et al (2000) has given the composition of minerals in finger millet referring to 7 elements. Shayo et al (2001) in Tanzanian 5 varieties reported calcium and phosphorous which are present at 10 to 340 mg% and 120 to 450 mg% phosphorous levels respectively. Surprisingly, phosphorous content is higher than calcium in these varieties. Majumdar et al (2006) have studied 3 hilly varieties and eight base varieties for calcium and iron content. In hilly varieties calcium ranged from 350 to 366 mg% while in base varieties it was 350 to 412 mg%. Iron in hilly varieties was present at 17.1 to 18.5 mg% and 11.3 to 16.8 mg%. Relatively, calcium levels were higher in most of the base varieties and iron was found to be high in hilly varieties. Indaf-11, a white variety had highest calcium which is comparable to earlier study. Shashi et al (2007) have reported the mineral content in 8 varieties. In the varieties of Indaf-5 and GPU28, the calcium and iron levels are lower as compared to Majumdar et al (2006). Secondly, potassium levels are highest and sodium levels are lowest as compared to other studies who have reported these elements. According to Glew et al (2008) Nigerian black variety had good calcium and iron levels at 401 and 18.2 mg% respectively. Ethiopian 9 varieties studied by Shimelis et al (2009) shows that calcium levels ranged from 50.6 to 319 mg% with iron levels of 4.59 to 53.39 mg%. Amongst the varieties studied so far by various researchers, Ethiopian PBR1 and PBR2 had the highest iron content of 53 mg% as well as the lower calcium variety Boneya with 50 mg% and Bambare and Ulyo Tanzanian varieties have lowest calcium of 10 and 30 mg% (Shayo et al, 2001). Radhika and Sarita (2008) have reported that amber (light coloured genotypes of finger millet had higher contents of calcium and iron than the dark genotypes. Shobana and Malleshi (2007) reported 180 mg% calcium and 109 mg% phosphorous in decorticated millet and as expected the levels are low in decorticated samples as it excludes the outer layer. Desai et al (2010) have analysed Indaf15 for calcium, iron and phosphorous content while Usha and Malleshi (2011) have also reported zinc and copper along with the above three common elements.

Thus over the six decades, the cultivated varieties have been studied for the minerals and the study encompassed the analysis of 9 elements i.e. Ca, P, Fe, Mg, Mn, Zn, Cu, K and Na. The marginal differences may be referred to the method of analysis. But the large differences and notable reports on particular element reflects clearly the varietal influence as well as agronomical conditions.

5. Vitamins

Fingermillet contain small quantities of some of the B series vitamins and the studies are limited. The reported literature refers to earlier years and the present researchers are not concentrating on this aspect. The earliest study reported vitamin B1 420 μg%, nicotinic acid 1100 μg%, and also carotene as 70 vitamin A units. Passmore and Sundararajan (1941) have studied 10 varieties and reported 0.27 to 0.65 mg% thiamine while Kadkol and

Swaminathan (1954) reported the range of 0.57 to 0.67 mg% thiamine in 8 varieties. Chitre et al (1955) in CO_2 variety studied have reported thiamine 0.19 mg%, riboflavin 2.5 mg% and niacin 0.1mg% ad this is the first report on riboflavin presence in finger millet. Mahadeshwaran and Ayyaperumal (1970) in their study on 11 white varieties and 9 brown varieties have reported 0.46 to 0.57 mg% thiamine, 0.1 to 0.15 mg% niacin and 0.11 to 0.22 mg% riboflavin. Deosthale et al (1970) have reported thiamine, niacin and riboflavin in ten varieties. Niacin was higher (0.27 to 0.87 mg %) followed by thiamine, 0.11 to 0.61 mg% and riboflavin, 0.02 to 0.07 mg%. They found that white variety has highest thiamine. Kamalanathan et al (1971) in their study on C09 white variety and local brown variety found thiamine at 0.41 mg% level and was almost same in both the varieties. Babu (1976) has studied in 6 varieties of fingermillet for folic acid content. Malleshi and Klopfenstein (1998) during their study on malting of fingermillet have reported 0.229 mg% thiamine and 0.093 mg% riboflavin, 2.22 mg% niacin, 0.11 mg% ascorbic acid in Indaf cultivar and is the first report of ascorbic acid in fingermillet.

CONCLUSION

The reviewed literature covering the global scenario on nutrients in 76 varieties of finger millet revealed that the composition of the grain comprise of the major nutrients as carbohydrates from 61 to 88.9%, protein 5.23 to 12.8% and lipids 1 to 4.5% reflecting the varietal difference probably due to the hybridisation, cultivation environment and practices. The composition of carbohydrates though contributes mainly to starch content, also has considerable quantum of 15 to 20% unavailable carbohydrates as dietary fibre component and small quantities of 2 to 4.5% free sugars. The starch component with 50 to 60% has 14.2 to 17.9% amylose and 82.1 to 86.1% amylopectin fractions and other starch fractions of nutritional importance have been dealt in Chapter 6. The major important nutrient protein, from varieties starting from 1954, accounting to 136 varieties ranged from 4.5 to 12.8%. Though new hybrid varieties are cultivated, the impact is not much on protein levels. Amongst the protein fractions, prolamine also known as elusinin is found to be one of the major fractions and others being albumin, globulin, glutelin which marginally varies with the varietal difference. But, the prolamines are similar to those of prolamines of maize, sorghum and teff. The amino acid pattern of finger millet protein as reported by many researchers reflects that it contains majority of the essential amino acids except the low levels of lysine and threonine. It is a good source of glutamic acid and the lysine present satisfies 50% of WHO recommendations. Amongst 121 varieties studied so far, the lipid levels range from 0.7 to 4.6%. The lipid classes were normal as neutral, glycol and phospholipids with slight lower content of phospholipids, but, followed similar pattern in all the varieties. However, the quantity varied very marginally amongst varieties, but, not significant. Fatty acid composition of finger millet as studied more widely compared to lipid classes have plamitic, stearic, oleic, linoleic and linolenic acids as common acids with others in small quantities reflecting 21 to30% essential fatty acids. Study on the hilly varieties' lipids reported is the first one and is from DFRL, Mysore.

Finger millet is a good source of minerals as minor nutrients and particularly calcium and iron has been studied exhaustively. Amongst 223 varietal samples studied for minerals, the broader range of calcium was 200 to 1116 mg% with an exception of lowest 10 mg% calcium in Tanzanian variety. Iron was found to be in the range of 1.3 to 37 mg% except highest 53 mg% iron in the Ethiopian variety. The other elements reported covers phosphorous, potassium, magnesium at considerable levels and sodium, zinc, manganese, copper at low levels. The earlier work of 1920-30's reported iodine content of 101 μg per kg which is highest among food grains. Calcium content is high in white varieties and iron is found higher in brown varieties. Besides this, the other observation is that calcium is high in base level grown brown varieties, while, iron is high in hilly grown varieties.

Finger millet contains small quantities of B-series vitamin and the studies confine to earlier years up to 1970's. Among 59 varieties studied, thiamine has the range from 0.19 to 0.67 mg%, riboflavin 0.02 to 0.25mg% and niacin 0.1 to 2.2mg%. Vitamin A 70 units and 0.11mg% vitamin C has also been reported to be present in finger millet. Considering the low levels of vitamins and expected further losses during processing, serious investigations on vitamins profile has not been dealt by the researchers.

Thus, over the six decades of research on minerals reflect the large differences and notable reports of particular element which is expected to be due to varietal influence as well as agronomical conditions of cultivation.

In: Finger Millet: A Valued Cereal
Editor: K.S. Premavalli

ISBN: 978-1-62081-224-2

Chapter 4

NON-NUTRIENTS OF FINGER MILLET

K. S. Premavalli and C. R. Vasudhish
Food Preservation Division, Defence Food Research Laboratory,
Siddharthanagar, Mysore, Karnataka, India

INTRODUCTION

Nutritional potential of finger millet in terms of protein, carbohydrate and lipids are comparable to the staple cereals in practice such as rice, wheat, barley, bajra. Each cereal apart from the nutritional strength has some deficiencies of nutrients as well as anti-nutritive factors. But, the type and quantum of undesirable factors are of paramount importance. Likewise, finger millet also contain phytates, tannins, oxalates, trypsin inhibitory factors, phenols and dietary fibre which are being considered as anti-nutrients due to their metal chelating and enzyme inhibiting properties. But, the present day researchers application on functional ingredients and properties reveal that these antinutrient factors with the specificity of structure, the form of occurrence show a positive behaviour on health benefits, in terms of antioxidant activity, aids in reducing cholesterol and act on better physiological actions. Therefore, these compounds can no longer be antinutrient factors, rather can be called as non-nutrients. Thus, "Non-nutrients may be defined as the compounds which do not contribute to the nutritional strength, but, exerts its both positive and non-positive action in the human metabolism at varied levels resulting in corresponding health benefits or inhibitory effects". Generally, these non-nutrients are present in small quantities except dietary fibre. The seed coat of the millet is an edible component of the kernel and is a rich source of phyto-chemicals such as polyphenols and seed coat extract showed high antibacterial and antifungal activity (Viswanath et al., 2009). Finger millet with dark brown or reddish brown has higher tannin levels as compared to dull white varieties with traces or no tannins (Geeta et al., 1977). All cereal grains are known to be rich in phytic acid (Reddy et al., 1982), which binds certain minerals, particularly divalent ions, thus making these biologically less available (Reinhold et al., 1975). Tannins complex with proteins and enzyme inhibitors and reduce digestibility (Reddy and Pierson, 1994). The increased antioxidant activity in free phenolic acids has been reported (Subba Rao and Muralikrishna, 2002). The dietary fibre in finger millet varieties has been reported by Premavalli et al (2004) and the health benefits associated with high fibre foods has been discussed (Tharanathan and Mahadevamma, 2003; Premavalli and Roopa, 2006). The processing methods adopted for preparation of finger millet based products have an impact on these non-nutrients thereby many a time it becomes a beneficiary step. Therefore, this chapter deals with the differential properties of non-nutrients.

1. PHYTATES

Phytates are naturally occurring phosphorous compounds which influences the functional and nutritional properties of foods. Though it is the source of phosphorous, phytic acid should not be regarded as an available source of phosphorous for humans or animals (Oberleas, 1973). The binding capacity of phytic acid readily forms complexes with multivalent cat-ions and proteins which are mostly insoluble in physiological pH thereby the bound minerals will be biologically unavailable. Phytic acid is present in the seed coats and hence the phytic acid content in cereals depends to a large extent on the milling process of grain and the extent to which bran and germ are separated from the endosperm. The phytates are decomposed to inositol and phosphoric acid by the enzyme phytase present in the grain. But, the phytase activity is relatively very low of 0.01 order for finger millet as compared to other cereals (Giri, 1938) and the author has reported 172 mg% phytin phosphorous in finger millet. Sankara Rao and Deosthale (1983) have reported an average value of 132 ± 2.3mg% phytate in 6 varieties of finger millet. Udayasekhara Rao and Deosthale (1988) have studied 12 brown varieties and 3 white varieties of finger millet and reported an average value of

188±7.5mg% in brown samples with 154 ± 6 mg% in white varieties reflecting the low levels of phytates in white varieties. Sri Lankan varieties CO10, KM1 and M1302 had higher level of phytic acid as 450, 490 and 490 mg% respectively (Ravindran, 1991). The common foods of tropical origin were analysed for phytate phosphorous and finger millet contain 140 mg% (Ravindran et al., 1994). Sripriya et al (1997) have reported 145 mg% phytate contents in finger millet while studying for the processing effect. Antony and Chandra (1998) in the variety used for their study, the phytate content was 510mg%. According to Mbithi et al (2000), the phytate present in the variety studied was 360 mg%. The phytic acid content in Kenyan varieties of finger millet has been found to be high. Makokha et al. (2002) have reported a range of 851.6 to 1419.4 mg% in five varieties. But, the Kenyan variety reported by Onyango et al., (2005) had low levels of phytic acid as 247 mg%. Wadikar et al., (2006) have reported phytate content in 3 hilly varieties and 7 base varieties of India. Relatively, on average, hilly varieties had slightly higher levels. It ranged from 212.3 to 241.4 mg% in hilly varieties and 115.9 to 262.6 mg% in base varieties with 248.9 mg% in white variety. The local Mandya variety called MRI had the lowest phytate of 115.9mg%. Shashi et al., (2007) in their study on 8 varieties reported phytate content of 171.7 to 320 mg%. The available literature clearly shows that the phytate phosphorous levels ranged from 115.9 to 1419.4 mg%, nearly 10 times higher and such a broad range reflects on the varietal difference and country of cultivation.

The processing of finger millet such as cooking germination, fermentation, hydrothermal processing, HTST methods of puffing and extrusion, known to reduce the phytate levels. However, the initial phytate levels of the grain have a greater impact on the utilisation for product development. During cooking, phytate content reduced by 12.8% (Giri, 1938). Phytate degradation during traditional cooking has been studied by Agte et al., (1999) in products of rice, wheat and all millets. The phytate degradation was highest with wheat and finger millet products. During germination and malting process, the significant reductions in phytates have been reported. Sankara Rao and Deosthale (1983) observed that malting of the grain significantly reduced the phytin phosphorous by about 36%. Udayasekhara Rao and Deosthale (1988) have shown the reduction of phytate with increasing period of germination upto 72 hours in both brown and white varieties. Irrespective of the varieties, phytate decreased by about 66%. According to Mbithi et al (2000), germination upto 96 hrs resulted in 94.5% reduction of phytates. The germination of finger millet upto 24 hrs followed by fermentation to a maximum of 18 hrs reduced the phytate content to the maximum (Sripriya et al., 1997). But fermentation of finger millet flour upto 24 hrs reduced phytate by 20% (Antony and Chandra, 1998). Mamiro et al., (2001) have showed that the combination of germination, autoclaving and fermentation reduces phytates by 85%. In 4 Kenyan varieties, Makokha et al., (2002) have reported the increased reduction of phytates with increasing fermentation time. After 96 hrs of fermentation, phytic acid reduced by 72.3% while malting upto 96 hrs reduced phytic acid by 45.3% and in all the cases the reduction was dependent on the varieties. The process of puffing resulted in 23.4 to 42.1% reduction of phytate in hilly varieties with 5.1 to 49.6% reduction in base varieties (Wadikar et al., 2006). The rate of reduction was dependent on the varieties and the white variety Indaf11 which had highest levels showed least decrease of 5.1% phytate. The extrusion cooking did not have an effect on degradation of phytic acid (Onyango et al., 2005). Thus, the processing method influences the degradation of phytates thereby reduction and it has varied from 0 to 94.5% reduction. Germination is one of the most effective methods to reduce phytates and the activation of endogenous phytases during germination is the causative factor. However, the varietal difference is of great concern in reduction of phytates.

The processing of finger millet inturn reduction in phytates is generally accompanied by improved bioavailability of the minerals. Since Calcium and iron are at higher levels many studies can be seen with reference to ionisable iron, calcium and their bioavailability. The ionisable iron expressed as percent of total iron, is an index of bioavailability of food iron (Narasinga Rao and Prabhavathi, 1978). The HCL extractability of minerals under simulated gastric conditions is also an indicator of availability from foods and thus generally followed. Sankara Rao and Deosthale (1983) had shown that ionisable iron increased from 7.4 to 88.3% during malting process, while, soluble zinc was relatively lower with an increase by 10% as compared to native grain. Udayasekhara Rao and Deosthale (1988) have found that in coloured ragi, 13% of iron was ionisable while in white varieties it was 22%, thereby, the bioavailability of iron in white ragi was significantly higher. But in the case of zinc, there was no significant difference between the coloured and white varieties. On germination, the available iron and zinc progressively increased with the period. But, in coloured varieties 72% of ionisable iron was available after 96 hrs germination while in white varieties it was only 56%. The changes were significantly correlated with phytate content in the coloured ragi. While, the germination did not show much difference in soluble zinc amongst varieties, but after 72 hrs the availability increased to 82 to 88%. Sripriya et al., (1997) have seen that germination of grains increased the extractability of copper, zinc, manganese which further increased on fermentation. Fermentation was most effective in increasing the bioavailability of calcium, phosphorous and iron by 13, 20, 15.5% respectively after 48 hrs fermentation. Antony and Chandra (1998) have showed an increase in HCl mineral extractability when fermented with endogenous grain microflora and reported increase of calcium extractability by 20%, iron 27%, zinc 26%, copper 78% and manganese 10%. According to Mamiro et al., (2001) the invitro extractability of calcium, iron and zinc significantly increased during germination while soaking, autoclaving and fermentation showed lesser effect. However, the addition of vitamin C increased the extractability by 8%. Gowri et al., (2001) have also reported the addition of amla, a good source of vitamin C in test meals enhanced the iron bioavailability. In Kenyan varieties of finger millet, Makokha et al., (2002) have reported that fermentation increased the rate of available iron, manganese and calcium. Shashi et al., (2007) reported that bioaccessibility of iron of finger millet genotypes ranges from 10 to 12mg% which was significant. Thus, the studies on bioavailability of minerals in finger millet are rather limited, but surely the processing methods adopted improve the bioavailability.

2. Phenolic Compounds

The natural colour of finger millet is because of phenolic compounds present in the seed coat. The main phenolic compounds in finger millet are phenolic acids and tannins with small quantities of flavonoids (Subba Rao and Muralikrishna, 2002). Finger millet varieties are coloured from brown to reddish brown to dark brown and dull white. The colour is mainly due to tannins and white variety has traces of tannins or it may not be present. Since the whole grain flour is used for preparation of products, the presence of tannins influence the sensory appeal as well as other characteristics. Secondly, tannins bind to proteins, carbohydrates and minerals, and thus reduce the digestibility of these nutrients. However, the processing method adopted influence in reducing these negative effects. On the other side, these compounds have functional properties such as antioxidant (Sripriya et al, 1996), antimutagenic, anti-oestrogenic, anti-inflammatory, antiviral effects and platelet aggregation inhibitory activity that might be beneficial in preventing diseases (Ferugson, 2001). Phenolic compounds in grains exist as free, soluble conjugates and insoluble bound forms. Hilu et al (1978) have reported that phenolic compounds are present in the form of glycosides while Subba Rao and Muralikrishna (2002) have reported the bound acid form. Finger millet phenolics are heat stable, but, pH sensitive and are unstable under alkaline conditions. Therefore, understanding of the type and quantum of phenolic compounds and their effect while processing are of importance to gain more of positive effects. Geeta et al (1977) studied 32 varieties including African varieties and hybrids with varied coloured grain. The tannins, i.e. polyphenolics as catechin were present at 0.03 to 0.06% in white varieties, while 0.52 to 2.02% in brown varieties and 0.14 to 1.05% in dark brown varieties, thus showing higher tannins in brown coloured grains. Udayasekhara Rao and Deosthale (1988) have also reported lower levels of tannins in dark brown varieties and higher levels in brown varieties. Amongst 12 varieties, tannin content varied from 0.35 to 2.39% with an average of 0.914 ± 0.014%. Shankara (1991) has reported tannins in 85 finger millet varieties which ranged from 0.19 to 3.37% and the difference amongst the varieties was due to the presence of the red pigments anthocyanins. Tannin contents from Eastern Africa vary widely from 0.27 to 2% (Mbithi et al, 2000). Wadikar et al (2006) have reported tannins in 3 hilly varieties and 7 base varieties. In hilly varieties it was 0.32 to 0.36% with 0.23 to 0.62% in base varieties. The hilly varieties had lower level of tannins as compared to base varieties. Shashi et al (2007) have reported 0.3 to 0.54% tannins in 8 finger millet genotypes. Thus finger millet varieties showed varied tannin levels from 0.2 to 3.5% tannins.

Dykes and Rooney (2006) have reported that finger millet also contain proanthocyanidins called condensed tannins which are high molecular weight polyphenols. But the structural characterization of millet proanthocyanidins has not been attempted. However, the condensed tannins are biologically active and when present in considerable quantities, may lower the nutritional value and biological availability of proteins and minerals (Chavan et al, 2001). Condensed tannins are generally more potent antioxidants than their corresponding monomers. Geeta et al (1977) have reported total phenols content in 32 varieties of brown and white varieties. Total phenols ranged from 0.053 to 2.44% in brown grains while dark brown varieties showed lesser phenols ranging from 0.37 to 0.96% and white varieties showed 0.06 to 0.1%. Antony and Chandra (1998) reported total phenols of 0.52% in finger millet, while Sripriya et al (1997) have reported 1.43% phenols in the variety studied. Phenolic acids may be present in free as well as bound form. Subba Rao and Muralikrishna (2002) have reported ferulic acid as the major bound phenolic acid and protocatechuic acid as the major free phenolic acid. Chethan et al (2007) have reported that the polyphenols in finger millet were the derivatives of benzoic acid and cinnamic acid and a flavonoid compound, quercitin. These acids were gallic acid, protocatechuic acid, p-hydroxy benzoic acid, p-coumaric acid, syringic acid, ferulic acid and transcinnamic acid. Shobana et al (2009) have found flavonols, glycosides, anthocyanins and condensed tannin in the seed coat extract.

The processing of finger millet in terms of milling cooking, decortication, germination, malting, fermentation, puffing, extrusion etc. have an influence on the phenolic compounds degradation and digestibility of nutrients. Geeta et al (1977) have reported on 13 varieties of finger millet in vitro protein digestibility (IVPD) variation with the tannin levels. The high tannin and phenol varieties showed lower values of IVPD as 55.4 to 57.4% and all other varieties including white variety showed 78 to 88% IVPD. There was no clear cut demarcation with reference to colour or tannins. The added tannic acid decreased IVPD by 40% from 83-43% and the effect increased with increase in tannic acid. The glutaline fraction had the highest tannin content which is likely to affect the digestibility thereby the nutritive value of the protein in finger millet. Germination process has decreased tannins and was progressively decreased with the time. The tannin content of wet grains steeped for 18 hrs was reduced by 20, 45, 62 and 72% with the germination for 0, 24, 48 and 72 hrs respectively (Udayasekhara Rao and Deosthale, 1988). Available ionizable iron increased with the germination progression and was greater in coloured than in white varieties.

The effect of fermentation has been reported by Antony and Chandra (1998) which reduces tannins by 52% and phenols by 20% at the end of 24 hrs and later the rate of loss was slow. IVPD increased from 47 to 70% with the fermentation time from 0 to 24 hrs and tannins, phenols were negatively correlated with IVPD. The same authors have found improved IVPD in different varieties (Antony et al, 1996, Antony and Chandra, 1999). Sripriya et al (1997) have studied the effect of germination and fermentation. Total phenols decreased on germination from 1.43 to 1.28%, but increased on fermentation to 1.86% which may be due to the hydrolysis of condensed tannins to lower molecular weight phenols (Khetarpaul and Chauhan, 1991). Accordingly, germination was more effective than fermentation step in increasing the extractability of trace elements, copper, zinc, manganese. But fermentation was more effective with respect to calcium, phosphorous, iron extractability. However, the combination of both processes showed a better potential for improvement of nutritional strength. On germination of finger millet, the inverse relationship between tannin content and IVPD has been reported by Mbithi et al (2000). The same authors have reported during processing of complimentary food by germination, autoclaving and fermentation reduced tannins to undetectable limits and IVPD increased to 90.2% from 69.5%. The effect of malting of finger millet has been studied by Subba Rao and Muralikrishna (2001, 2002). The changes in free bound and free phenolic acids during malting for 96 hrs showed a decrease of bound caffeic, coumaric and ferulic acid levels by 45, 41 and 48% respectively whereas free gallic, vanillic, coumaric and ferulic acids increased considerably. Cooking of millet flours reduced the total phenols (Matuschek et al, 2001). The extrusion process reduced tannins from 1.67 to 0.697% while further fermentation decreased the level to 0.55%. But the presence of acids showed a lesser effect (Onyango et al, 2005). Puffing of finger millet reduced tannins, but was dependent not only on the variety but also the altitude level of cultivation (Wadikar et al, 2006). They have found that in hilly varieties, the reduction was only 2.8 to 3.7% which may be due to intactness in seed coat because of low temperature of cultivation. While, in base varieties, tannins reduced from 13-22% and the white variety Indaf11 showed 13% loss. Chethan et al (2008b) have reported about 44% of loss of polyphenols during 24 h germination and increased to about 80% after 120 h of germination. The only millet flavonoids reported by Hilu et al (1978) identified eight flavones in the leaves of finger millet, namely orientin, isoorientin, vitexin, isovitexin, saponarin, violauthin, lucenin-1 and tricin.

3. Inhibitors and Oxalates

The presence of enzyme inhibitors decrease the activity of enzymes thereby the digestion and absorption gets affected. The polyphenols inhibit the activity of digestive enzymes such as amylase, glucosidase, pepsin, trypsin and lipases (Rhon et al 2002). Antony and Chandra (1998) reported that during fermentation of finger millet, trypsin inhibitor activity reduced by 32%. As reported by Mbithi et al (2002) during processing of complimentary food with finger millet by germination, autoclaving and fermentation trypsin inhibitor was reduced to undetectable limits. Chethan et al (2008b) studied the mode of inhibition of finger millet amylases by the millet phenols and amongst the phenolic acid trans cinnamic acid showed a higher degree of inhibition as 79%. Rhon et al (2002) have shown that decrease in enzyme activity depends on the concentration and type of phenols. The content of total oxalate was found to be low in finger millet. Ravindran (1991) has reported, 29-30% total oxalates in three varieties, but, 50% was water soluble form which are expected to be leached out during normal cooking (Libert and Franceschi,

1987). The same level of oxalates occur in many fruits and vegetables and do not pose a nutritional problem (Fasset, 1973).

4. DIETARY FIBRE

Finger millet carbohydrates comprises of starch as the main component with non-starchy polysaccharides accounting 15-20% of the seed as an unavailable carbohydrate which is regarded as dietary fibre which escapes the digestion in the intestines but still remain as an important component of our diet. Dietary fibre is mainly classified into soluble and insoluble fractions and both have their own specific roles in the physiological actions. Dietary fibres are considered as multifunctional substances having beneficial effects against several diseases such as, cardiovascular problems, diverticulosis, diabetes mellitus, colon cancer, appendicitis, constipation, hemorrhoids, hernia, duodenal ulcer, gall stone, obesity and other diseases of gastrointestinal tract (Slavin, 2005; Gomez et al, 2010). Insoluble fibre contains cellulose, lignin, hemicelluloses which has an impact on gastrointestinal transit time, binds water, speeds up intestinal transit, increase fecal bulk and binds some carcinogens. Soluble fibre encompasses pectins, gums and some hemicelluloses which absorb water, gets fermented in the large intestine microflora thereby bring out desirable metabolic effects (Lopez et al, 1999) soluble fibre has the ability to fatty substances in gastrointestinal tract thereby reduce cholesterol in the blood which impart better cardiac health. Soluble fibre also retard absorption of glucose in the small intestine thereby reduces the blood sugar level (Onyango et al, 2004b). The in vitro studies on functional properties of natural vegetable fibres as reported by Prachi Gupta and Premavalli (2011) infers that the fibres studied showed an excellent performance of hypoglycemic effect. Dietary fibre with phenolics as a synergist may play a role in inhibitory amylases thereby contribute to the management of type II diabetes (Toeller 1994). Soluble fibre though it is not a nutrient, the increased soluble fibre content has a special nutritional significance because of physiological benefits in terms of hypoglycemic and hypocholesterolemic characteristics (Shobana and Malleshi, 2007). Some fibres can absorb mutagenic agents and are eliminated in the faeces (Thebaudin et al, 1997). At the same time the dietary fibre reduces the absorption of vitamins, minerals and fermentation in the large intestine, form gases which may cause flatulence problems. Dietary fibres of finger millet contributes to 17-20% and the major portion is the insoluble fibre fraction.

Navitha and Sumathi (1992) have analysed the fibre profile of ragi and reported the total dietary fibre as 17.6% with 15.7% insoluble fibre and 1.8% soluble fibre. Besides these, neutral detergent fibre which constitutes to the components such as hemicelluloses, lignin, cellulose and cutin, silica have been reported by Thomas et al (1990) as well as Navitha and Sumathi (1992). This fraction comprise mainly of insoluble fibre and has been reported as 15.6% in finger millet. Premavalli et al (2004) have studied the dietary fibre profile of 10 varieties of finger millet grown at varied environmental conditions. Three hilly varieties reported had the total dietary fibre range of 15.5 – 17.8% while, amongst 7 base varieties, all the varieties showed 17.8 to 20% except the white variety, Indaf 11 with 13.8%. Thus the hilly varieties had relatively lower dietary fibre than the base varieties and white variety has lowest dietary fibre (Table 4.1). Finger millet varieties showed the major fraction of insoluble dietary fibre ranging 14.88 to 17.05% in hilly varieties and from 17.07 to 19.22% in base brown varieties with 13.21 in white variety. The soluble fibre fraction was found to be low recording 0.61 to 0.78% in hilly samples while, 0.72 to 0.89% in base samples, with 0.6% in white variety. The changes in fibre profile amongst the varieties was statistically significant. Usha and Malleshi (2011) have reported 17.1% dietary fibre in GPU-28 variety with 15.7% insoluble and 1.4% soluble fractions.

Table 4.1. Dietary Fibre Profile of Finger Millet Varieties

Varieties	Dietary fibre, %		
	Insoluble	Soluble	Total
5500 ft asl			
VL146	14.88	0.61	15.49
VL149	16.22	0.65	16.53
VL204	17.05	0.78	17.82
2300 ft asl			
Indaf5	18.76	0.76	19.90
Indaf8	18.15	0.89	19.04
Indaf9	17.07	0.72	17.8
Indaf11	13.21	0.60	13.8
GPU28	18.25	0.77	19.1
MR1	18.05	0.74	19.84
HR911	19.22	0.78	20.03
Mean±SD	17.08 ± 1.8	0.73 ± 0.08	17.93 ± 2

Premavalli et al, (2004).

Table 4.2. Effect of Processing on Dietary Fiber Fractions* (%) in Control System(n=10)

Processing	Insoluble fiber	Soluble fiber	Total fiber
Native flour	18.10	0.74	19.84
Cooking	12.97 ± 0.2	1.91 ± 0.2	14.88 ± 0.2
Pressure cooking	14.03 ± 0.2	1.62 ± 0.1	15.65 ± 0.5
Autoclaving	13.93 ± 0.2	1.51 ± 0.1	15.44 ± 0.1
Re-autoclaving	12.75 ± 0.1	1.90 ± 0.1	14.65 ± 0.2
Puffing	19.56 ± 0.2	0.76 ± 0.2	20.32 ± 0.2
Roasting	13.10 ± 0.4	1.60 ± 0.2	14.70 ± 0.1
Baking	8.55 ± 0.2	1.15 ± 0.3	9.70 ± 0.4
Frying	11.06 ± 0.9	1.10 ± 0.3	12.16 ± 1.0
Germination	8.86 ± 0.3	1.80 ± 0.1	10.66 ± 0.4
Malting	8.75 ± 0.1	3.25 ± 0.1	12.00 ± 0.4
Toasting (1)	12.57 ± 0.2	1.05 ± 0.3	13.63 ± 0.3
Toasting (2)	9.75 ± 0.1	1.34 ± 0.1	11.09 ± 0.2

*Mean + SD.

Roopa and Premavalli (2008).

The processing of finger millet and its products preparation will have an influence on the characteristics properties of the grain. The various processing methods decreased the total dietary fibre by 21-49% except during puffing it increased by 2.5% (Table 4.2). Similar pattern was observed in the changes regarding insoluble fraction. However, the soluble fibre fraction increased in all the methods of processing and the highest increase was during malting while the lowest increase was during puffing of finger millet. Increased soluble fibre fraction is associated with solubilisation of some insoluble fibre fraction, disruption of ligno-cellulosic links in the cell walls of fibre into low molecular weight soluble fragments (Bjork et al, 1984; Periago et al, 1996). Solubilisation of fibre makes it easier to ferment in the colon. The decrease in total dietary fibre fraction during extrusion of unfermented maize-finger millet has been reported by Onyango et al (2004b). But extrusion in acidic conditions increase the insoluble fraction because of the polymerization of short chain fibre fragments to form larger insoluble complexes (Camire, 2001). Probably, the similar mechanism holds good for the increased insoluble fibre fraction during puffing. According to Usha and Malleshi (2011), hydrothermal processing did not show much influence on total, insoluble dietary fibre but decortication has reduced total and insoluble dietary fibres. They have reported 16.9 and 16% total and insoluble fibre fractions respectively, during hydrothermal processing, while, it was further reduced to 10.1 and 7.8% in decorticated samples which may be due to the seed coat exclusion during decortications. But the soluble fibre decreased during hydrothermal processing by 36%, while increased during decortication by 64% and the changes have been related to increased water soluble and decreased cellulosic fractions. The primary processing of milling had shown highest total dietary fibre in finger millet as compared to other cereals (Navitha and Sumathi 1992). Processing of most of the cereals namely wheat, sorghum, bajra, maize, ragi in to chapathies had no effect on fibre profile except ragi where a significant increase in total dietary fibre and insoluble dietary fibre have been observed and this increase could be due to resistance of tannin bound proteins to enzymic hydrolysis of proteins (Ramulu and Udayasekhara Rao, 1997). In cereal based food preparations, Sharavathy et al (2001) have reported higher dietary fibre in ragi roti as compared to other foods.

Resistant starch called the 'functional fibre' is present in finger millet which escapes the enzymatic digestion and is found at low levels of 0.56 to 0.77% (Roopa and Premavalli, 2008). The changes during various processing methods have been discussed by many researchers. However, the details of resistant starch changes will be discussed in the chapter on finger millet starch and its properties.

5. Functional Properties of Non-Nutrients

Non-nutrients specifically phenolic compounds and dietary fibre exhibit functional properties imparting several health benefits. Sripriya et al (1996) have reported that finger millet is a potent source of antioxidative compounds and the possible involvement of phenols, tannins and phytate in the antioxidative effects have been discussed. Bravo (1998) discussed that antioxidant activity as an important factor in health, aging and metabolic diseases and these compounds play a role. Ferguson (2001) reported that polyphenols is one of the widest group of dietary supplements marketed worldwide. Chethan et al (2008a) have reported on the DPPH scavenging activity of crude phenolic extract as well as different phenolic acids revealing its antioxidant potential. Vishwanath et al (2009) shown the antioxidant activity of methanolic extract of seed coat. Subba Rao and Muralikrishna (2002) reported that the antioxidant activity of a free phenolic acid mixture was found to be higher than bound phenolics of finger millet. Ferulic acid exhibits strong antioxidant, free radical scavenging and anti inflammatory activity (Castelluccio et al, 1995; Shahidi et al, 1992).

Antony and Chandra (1998) in their study on fermentation with starters have shown that fermented finger millet extract suppress the growth of Salmonella and Escherichia coli. While, Chethan and Malleshi (2007) have studied the effect of germination on microbial activity which infers that millet phenols exert antimicrobial properties. Phenols are mainly present in the seed coat acts on the fungi and is the causative factor for long shelf life of grains (Viswanath et al, 2009). The enzyme inhibition on a positive role influence the diabetic condition. According to Gopalan (1981), regular consumption of finger millet is known to reduce the risk of diabetes mellitus and play a role in gastrointestinal tract disorders (Tovey, 1994). These properties were attributed to its high polyphenols and dietary fibre contents. The role of dietary fibre is attributed to slower gastric emptying or formation of complexes with available carbohydrates thereby delayed and less absorption of carbohydrates. Hegde et al (2002) have shown that the methanolic extracts reduces the complication of diabetes and aging due to the antioxidant properties and it prevents glycation and cross linking of collagen. Rajasekaran et al (2004) have supported the improvement of antioxidant status and provide the health benefits. Phenols are known to inhibit enzymes and the inhibition of amylases leads to the reduction of post prandial hyperglycemia (Chethan et al, 2008b) and inhibition of aldose reductase reduce the risk of diabetes induced cataract diseases (Chethan et al, 2008a). Shobana et al (2010) have showed that finger millet seed coat feeding has led to the reduction of diabetic complications. Oxidative stress and hyperglycemia in diabetes produce reactive oxygen species which causes peroxidation membrane lipids, protein glycation and health complications (Monnier, 1990). Antioxidants properties of millet reduces the peroxidation and these actions have been studied in rats (Chattopadhyay et al, 1997; Hegde et al, 2005). The antioxidant activity also prevent tissue damage and stimulate the wound healing process (King 2001; Rajasekaran et al, 2004).

Dietary fibre from finger millet because of high insoluble fibre, is associated with various health benefits in terms of increased faecal bulk, lowering of blood lipids, slow digestion, prevention of colon cancer, increased transit time and fermentation in the intestine. Theabaudin et al (1997) have reported that some fibres can absorb mutagenic agents which are later eliminated in the faeces. Resistance starch called the functional fibre is present in ragi which escapes the enzymatic digestion and provide benefits specifically to prevent colon cancer (Englyst et al 1992; Gee et al 1992). Resistant starch with their occurrence in foods has been reviewed by Premavalli and Roopa

(2006). Besides the health benefits, resistant starch also improve sensory properties. Thus the various non-nutrients present in finger millet besides having antinutritional properties, also provide functional impact to impart health benefits.

CONCLUSION

Like any other cereal, finger millet also possesses some deficiencies as well as antinutrient factors. But the type and quantum of antinutrient factors is of paramount importance. Finger millet contain phytates, tannins, oxalate, trypsin inhibitor, phenols and dietary fibre which are considered as antinutrients due to their metal chelating, protein binding and enzyme inhibiting properties. In 45 varieties reported phytates are present in Indian varieties in the normal range of 115.9 to 248 mg% except the high levels in Sri Lankan varieties, 470 to 490 mg% and Kenyan varieties 851 to1419 mg%. Such a broad range reflects on the varietal difference and country of cultivation. The various processing methods generally adopted during product development reduced the phytates from 5 to 94.5% depending on the method, in turn bioavailability of minerals improved by 7.4 to 88.3%. Phenolic compounds comprise of tannins phenolic acids and traces of flavonoids where tannins provide colour to the grain. In 147 varieties studied tannins ranged from 0.14 to 3.37% in coloured varieties and traces in white variety. While, 34 varieties studied phenol contents reported is 0.053 to 2.44%. The processing methods have shown the reduction of phenolics by about 50% maximum with improved invitro protein digestibility. The processing of grains also reduces or eliminates the trypsin inhibitors and the oxalates found in finger millet are as found in fruits and vegetables which do not pose any nutritional problem. Dietary fibre of finger millet contributes to 15-20% as studied in 23 varieties with insoluble fibre 14 to 19.2% and soluble fibre 0.6 to 0.89% and first time reported in hilly varieties is from the study of DFRL, Mysore. The processing influenced the dietary fibre profile recording decrease but increase in soluble fibre. The antioxidant potential of phenolic compounds and phytates, antimicrobial and enzymic inhibition of phenols has been reported. Dietary fibre and polyphenols have shown to impart several health benefits mainly reduce the risk of diabetes and its complications, cataract diseases, tissue damage, etc. Dietary fibre is also associated with increased faecal bulk, lowering of blood lipids, slow digestion, increased transit time and fermentation in the intestine. Resistant starch, a functional fibre prevents the colon cancer.

Thus, in this Chapter, a new nomenclature of 'non-nutrients' is coined for these compounds and "Non-nutrients may be defined as the compounds which do not contribute to the nutritional strength, but, exerts its both positive and non-positive action in the human metabolism at varied levels resulting in corresponding health benefits or inhibitory effects".

In: Finger Millet: A Valued Cereal
Editor: K.S. Premavalli
ISBN: 978-1-62081-224-2

Chapter 5

PROCESSING OF FINGER MILLET

K. S. Premavalli and C. V. Madhura
Food Preservation Division, Defence Food Research Laboratory, Siddharthanagar, Mysore, Karnataka, India

INTRODUCTION

The demand from the consumer calls for any agricultural produce from the land to undergo processing prior to utilization for any purpose further. Finger millet is no exception and as a small grain cereal requires more attention and the losses also may be high. The primary processing is adopted after harvesting by way of cleaning, destoning and storage. Secondly, cleaned grains are ground into flour i.e. milled flour for further use for the preparation of secondary processed products such as dumpling/thick porridge, a cooked product; roti, a toasted product; ambli/thin porridge, beverage, a cooked gruel and so on. The secondary processing such as germination, malting, fermentation, puffing, extrusion cooking, baking are being adopted for preparation of tertiary products which may be in ready-to-eat form or ready-to-reconstitute form or requires a little handling steps prior to consumption. However, the research on the optimization of the processes, changes that occur on characteristics, nutrients of finger millet grain, fate of antinutrients or enzyme inhibitors, formation of hydrolytic enzymes etc. are being investigated by many groups, to have an in-depth understanding of mechanisms. In this chapter, the various processing methods for finger millet and the scientific considerations are being discussed. Though finger millet is nutritionally strengthful, the colour, the small size, intact endosperm, nature of fat are some of the limitations for processing. But, the type of the starch, calcium and iron content, dietary fibre and the strong flavor generation during puffing makes it more advantageous and preferable grain for the health benefits.

1. MILLING

Milling is the primary processing adopted for utilization of finger millet *Direct milling* of grains through Roller milling result in flour where the endosperm and bran are pulverised freely. This flour will be rich in dietary fibre content and possess the natural colour of finger millet from reddish brown to dark brown. *Hydrothermal processing* of grains for milling of flour covers moistening, conditioning, drying, grinding and sieving. Generally finger millet is steeped for 10 hours, wet grains are steamed, dried to 12-13% moisture, ground into flour. Though hydrothermal processing improves milling efficiency, it enhances the hardness and aids in preparing grits. But this hardening will have an impact on the quality of the product. Further to hydrothermal treatment, *decortication* of the grains can be achieved prior to milling. This process results in refined flour decoloursied so that the limitation of dark colour of ragi can be overcome. Hydrothermally treated finger millet is passed through disc mill to remove the husk with less pressure. Thus obtained polished grains is dried and ground into flour for further use. Secondly, polished grains can be directly used for the preparation of idli, dosa, upma, cooked rice and extruded products. The polished grains will be harder than wheat grits. Thirdly, polished grains can be processed further into millet flakes by soaking, steaming, partial drying, processing through rollers into flakes and drying to low moisture for better stability.

Parboiling is generally followed for rice processing which imparts glazy appearance, better digestibility and maintains grainy structure after cooking. Desikachar (1972) has used parboiling or wet heat treatment for finger millet and reported that it enhances hardness. Katti et al (2008) have seen the effect of milling of finger millet in different pulverisers and the damaged starch content was highest in roller mills followed by emery plate mill and pin and hammer mill flours. The physical properties in terms of type of particles by SEM revealed that the flours from the plate, emery and roller mills had large number of starch granules exposed whereas pin and hammer milled flour had intact starch granules. X-ray diffraction pattern showed the decreased crystal size of starch with increased damaged starch content. The nutrient composition of raw finger millet flour i.e. directly milled flour in terms of carbohydrates (Premavalli et al, 2004), lipids (Wadikar et al, 2007), antinutrients (Wadikar et al, 2006) and minerals (Majumdar et al, 2006) have been reported in ten varieties of finger millet. A process for preparing decorticated finger millet has been patented by Malleshi (2003). The process involves cleaning, steeping for 2-6 hours at 20-70°C, steaming under pressure less than 5 Kg/cm2 for 2-20 mins, drying to 8-16% moisture, passing through disc mill and aspiration of seed coat to get decorticated millet which can be used as such or ground to utilize for other products preparation. Hydrothermal processing of finger millet has been described by Usha and Malleshi (2011). Finger millet was steeped in water at 30°C for about 10 hours. After discarding excess water, autoclaved at atmospheric pressure for 30 mins, dried at 38°C to 13% moisture. Hydrothermal processing though do not alter the gross composition of finger millet, the process has an impact on the major nutrients. Starch undergoes slight thermal degradation leading to an apparent increase in the amylase content. It also increased the carbohydrate and protein digestibility. Decortication of hydrothermally processed millet increased the soluble portions of dietary fibre and also non starch polysaccharide fractions. Though decortications lowers the nutrients may increase the bioavailability of minerals because of decreased dietary fibre.

2. MALTING

Malting of finger millet is being practiced in Southern India over the ages. Finger millet has superior malting properties compared to other cereal grains. Malting of finger millet does not pose problems of mold growth, off odour etc. Finger millet malt based products have acceptable taste, very good aroma and shelf life. *Malting* is generally followed by washing the good quality grain and steeping for about 18 hrs with changing the soaked water minimum twice. The soaked grains after washing, draining the excess water will be germinated by tieing in cloth for 36-48 hrs depending on the temperature and humidity prevailing. Generally, water to be sprinkled during germination to keep the sprouts moist. After the incubation period, sprouted grains are dried, rootlets eliminated and dried by mild toasting or kilning at 65-70°C in an iron pan under low flame. Then ground into flour and sieved. The refined flour blended with sugar and flavor result in an Indian traditional product called "*Ragi malt*" which can be consumed with milk. Malted flour can also be blended with malted or toasted legumes to produce weaning food and supplementary foods.

Besides these, toasted grains can also be used as an adjunct in brewing to produce malt extract and malt syrup.

Malted finger millet has been used for child feeding and as a thickener in the milk based beverages (Narayana Rao et al, 1958). During malting process, hydrolytic enzymes are formed and the varieties which form higher levels of enzymes show low malting loss thereby such varieties are considered good for malting (Chandrashekara and Swaminathan, 1953; Malleshi and Deiskachar, 1979; Nont and Davis, 1982). Though nutrient losses in terms of minor nutrients occurred during malting of finger millet, the bioavailability of minerals improved (Sankara Rao and Deosthale, 1983). Probably the reduction in phytates on malting may be the cause for increased bioavailability of minerals. Dada and Dendy (1988) have studied the effect of germination on cyanide content and found that both in ungerminated and germinated finger millet for 2 days, hydrogen cyanide was not detected. Finger millet malted by steeping for 16 hrs followed by germination for 48 hrs, drying, grinding and sieving was used at 30% level for the preparation of weaning food with mung beans (Malleshi and Desikachar, 1982). They have observed a significant increase in amylase activity and decrease in paste viscosity with progressive germination. Relatively finger millet showed higher increase as compared to green gram. Malleshi and Amla (1988) have conducted the shelf life studies of finger millet based weaning food and the supplementation programme with children. Mtebe et al (1993) have shown that germination of finger millet is useful in improving energy density, bioavailability of nutrients and acceptability of togwa, a traditional sweet gruel. The study of Sripriya et al (1997) reveals that germination of finger millet was effective in starch and protein hydrolysis. Malleshi and Klopfenstein (1998) have reported 78% as the yield of finger millet malt flour. The protein and crude fibre contents were 4.5% and 1.8% which are nearly half of the native flour. Mbithi Mwikya et al (2000) have studied the effect of germination upto 96 hrs period on nutritional parameters and reported improved protein digestibility, improved mineral extractability, increased sugars and decreased phytates during germination.

Nirmala et al (2000) showed that Indaf-15 finger millet variety has a good malting potential because of high levels of reducing sugars produced and high amylase activity during malting. Kalpana Platel et al (2010) have shown the influence of malting on the bioaccessibility of minerals in finger millet with reference to iron, zinc, calcium, copper and manganese and a comparative study on wheat and barley has been reported.

3. Fermentation

Fermentation is a process in which carbohydrates are broken down under aerobic or anaerobic conditions into alcohol and acids and is a process used to produce beverages such as wine, beer, cider. During the process, the action of microorganisms brings about a desirable change in flavor, digestibility, minor nutrients and functional components such as antioxidants, phenolics, and phytochemicals.

Generally, cereals are used to obtain the fermentation products such as idli, dosa, bhathura, most common in India, but globally, cereals of second line are used in brewing industry for the manufacture of alcoholic beverages. Preprocesses followed prior to brewing are germination and malting of grains. Finger millet is steeped for 16 hrs followed by 2 to 3 days germination, drying, and grinding into malted flour to use for brewing for the production of beer.

Fermentation techniques differing in starter culture, period and temperature have been followed along with varied characteristics of cereals. Lorri and Svanberg (1993a, 1993b) have studied the lactic fermentation of finger millet gruel using different starter cultures and the fermentation significantly improved in vitro protein digestibility. In the unfermented non-tannin finger millet gruel, the in vitro protein digestibility had higher values of 61 to 79% but lesser than the fermented products. Antony et al (1996) have studied the effect of fermentation on the primary nutrients in finger millet. The fermentation decreased the levels of starch, long chain fatty acids and pH, while, acetic and lactic acids were produced and increased total microbial flora was observed. The same group of workers further concluded that germination of finger millet was effective in starch and protein hydrolysis while fermentation using malted flour was effective in reducing the pH, phytate content with the increase in mineral extractability, free sugars and amino acids (Sripriya et al, 1997). Antony and Chandra (1997) declared that natural fermentation of finger millet flour resulted in decreased pH, increased acidity and the process was heterolactic dominated type. The microflora stabilized between 24 to 48 hrs and identified five predominant types of bacteria belonging to genera *Leuconostoc, Pediococcus* and *Lactobacillus*. At the end of fermentation, increased amylase activity, sugars and amino acids were observed. The same authors have studied the effects of fermentation of finger millet flour using endogenous grain microflora on antinutritional factors as well as primary nutrients. The study revealed the significant reduction of phytates, phenols, tannins and trypsin inhibitor as well as increased mineral extractability and in vitro protein and starch digestibility (Antony and Chandra, 1998).

The controlled fermentation by treatment with enzymes followed by fermentation achieved the desirable goals of reduced fermentation time to 12 hrs as compared to 48 hrs of uncontrolled natural fermentation. Enzymatic treatment did not significantly alter the pH, phytate, tannins and mineral extractability, but enhanced fermentative changes (Antony and Chandra, 1999). Jani et al (1999) conducted the studies on effects of germination time up to 5 days during malting on brewing properties of resulting malt. Pentosans and dynamic viscosity increased with the germination time. Onyango et al (2000) described the fermentation process for the production of uji and the conditions used were 45°C for 24 hrs. Anandhi and Rashmi Kapoor (2003) have seen the effect of natural and pure culture fermentation of finger millet on zinc availability. They found that pure culture fermentation that too Lactobacillus spp. was more effective than natural fermentation in increasing zinc availability. Hemalatha et al (2007) have reported that the germination of finger millet decreased the zinc bioaccessibility while improved iron bioaccessibility. But the combination of germination and fermentation enhanced the bioaccessibility of both zinc and iron during idli, dosa products preparation.

4. Puffing

Puffing also called popping is traditionally adopted to provide convenience for preparation of millet products. *Puffing* is followed by moistening with 20% water, aged for 2 hrs and the wet grains are puffed using hot sand at 230 to 270°C. Puffing is a high temperature short time process which results in gelatinization of starch and flavor generation. Cleaned, puffed grains is ground into flour for further use. Blending of puffed flour with malted or roasted legumes leads to several products viz. weaning foods, supplementary foods, health foods, ready-to-eat snacks and so on.

Malleshi and Desikachar (1981) optimized the conditions for puffing of finger millet by moistening finger millet grains to 19% moisture, attaining equilibrium by 4 hrs keeping, followed by puffing in sand medium at 270°C. They found variation in puffing quality among 14 varieties and Purna, Annapurna, Shakti, PR202, Indaf 3 showed good puffing quality. However, no consistent relationship was observed between amylase, proteins with puffing quality. Shukla et al (1986), Sarita and Anju (1998) reported popping quality of genotypes of finger millet and the popping ranged from 55.8 to 87.5%. But, the expansion volume was not related to puffing percentage. Premavalli et al (2005) have studied the effect of pretreatments on the puffing characteristics of finger millet. The effect of hydration of finger millet MRI variety was compared using water, butter milk and citric acid solution for 2 hours equilibration followed by puffing using hot sand at 260 to 270°C. It was also compared with pan roasting. The puffing yield was found to be higher in butter milk treated samples (96%) as compared to water (91%) citric acid (85%) and least in pan roasted samples (84%) and the swelling power and water absorption was directly related with the puffing. But the bulk density was highest for pan roasted sample than others; however, it was not significant. The higher water absorption index and swelling power indicate the gelatinized flour which is more susceptible for hydration. The concomitant changes in percent gelatinization is observed with the pretreatments of grain. The study led to a conclusion that water and butter milk are better hydration medium for puffing of finger millet, but water conditioning is more economical. The puffing did not result in any considerable change in chemical parameters. The effect of puffing on calcium and iron contents of finger millet (Majumdar et al, 2006), on antinutrients, (Wadikar et al 2006) on lipid profile (Wadikar et al, 2007) have been reported. However, the details will be covered in other respective chapters. The puffing yield of ten varieties of Ragi (Premavalli et al, 2006) ranged from 70 to 91% with lower puffing in hill grown varieties. The puffed grains profile is presented in Figure 5.1. Radhika and Sarita (2008) have reported the differential behavior of genotypes on popping characteristics. Popping varied with the variety and ranged from 54 to 84%. Popping and expansion volume also followed the same pattern and genotype PES-400 had maximum values of all these parameters and hence considered best in terms of popping characteristics. Malleshi and Singh (2005) has patented a process for preparation of expanded millet. Ushakumari et al (2007) have tried the possibility of preparing expanded millet from decorticated finger millet and optimized the process by response surface methodology.

5. Extrusion Cooking

The advancement in technological field has led to the extrusion cooking of foods which is an HTST Process. This process can aid in producing low energy foods such as low fat as well as low protein foods. Extrusion technology can be used for producing ready-to-eat snacks in terms of value added products with health benefits.

Finger millet has been used as a potential cereal for producing vermicelli. Sudha et al (1998) studied with 4 different particle size finger millet flour incorporated with wheat extruded vermicelli. from medium particle size finger millet flour (50%) with hot water dough preparation and extrusion showed higher sensory scores. Onyango et al (2005) attempted an industrial extrusion process for the production of uji, in a convenient dry form using finger millet based slurry and found that extrusion reduced tannins but not phytates in the product. The extruded products with finger millet flour may lead to colour change, hardness in texture and storable problems. But have the advantage of increased dietary fibre and may be economical. Therefore, pertaining to this aspect, it is open for further research.

6. Baking

Baking process is an age old technology followed for production of several bakery products. Generally blending of finger millet flour up to 30-40% with Maida/wheat flour is followed to produce finger millet products. Simple addition of finger millet flour into the formulation result in grainy, not cooked/bad feel of the product and is because of the different starch properties of finger millet flour and maida flour. Though baked products is most promising to have ready-to-eat snacks, modification of finger millet incorporation is required. In this direction, research carried out in our laboratory revealed that modification of starch by way of puffing or roasting the grain improves certain of the characteristics.

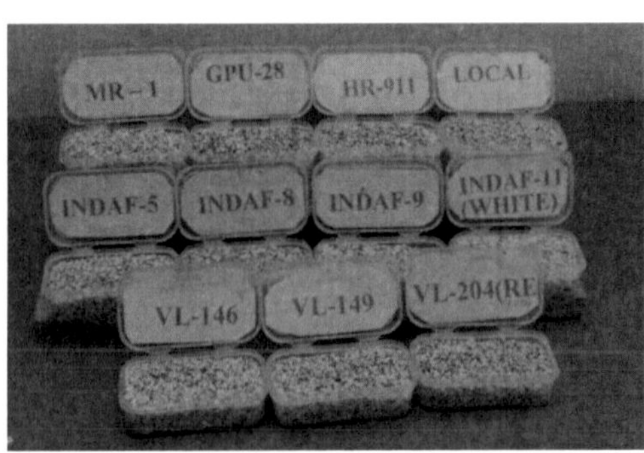

Figure 5.1. Finger Millet Puffed Grains Profile of Indian Variety.

7. Combination Processing

The combination of processes generally aids at synergistic effect so that one will be complimentary to the other. Mbithi et al (2000) have studied the effect of sprouting, autoclaving and fermentation on amino acids in turn protein quality for the preparation of weaning food for children. Besides the provision of good percentage of total amino acids, 44.2 to 44.9% in finger millet as compared to 33.9% for the FAO/WHO reference protein for 2-5 years old children, the processes showed significant changes in individual amino acids too. Autoclaving caused decrease in histidine, while fermentation decreased phenylalanine and increased tryptophan in finger millet. During sprouting and fermentation, leusine to lysine ratio, which is an indicator of the pellagragenic character of a protein was improved in finger millet. Mamiro et al (2001) have shown improved mineral extractability in germinated finger millet. Mamiro et al (2004) processed the finger millet by germination, autoclaving and drying for the weaning food with other legumes and reported a reduction of phytates by 34% and increased iron solubility by 19% during processing. The combination processing of germination and fermentation of finger millet has been assessed by Hemalatha et al (2007) and bioaccessibility of zinc decreased by 30% during germination while iron bioaccessibility increased by 20% and tannins decreased. However, fermentation did not improve the bioaccessibility.

The effect of cooking and spray drying of finger millet hydrolysate showed greater amorphous component than the native flour which has been explained as the breakage of weak hydrogen bonds that are essential for the maintenance of an orderly polymer net work and a perfect crystalline region.

The effect of processing i.e. cooking, autoclaving, puffing, roasting, baking, frying, toasting, germination and malting of finger millet flour on starch fractions has been studied by Roopa and Premavalli (2008). Readily digestible starch increased in all the processing methods except for baking, frying and toasting in oil. The details of the study will be discussed in the chapter Finger Millet Starch and its Properties.

Conclusion

Processing of grain is a necessity for improving digestibility to eliminate antinutrient factors, to generate flavour, taste and later for further utilisation and it becomes more prominent at the instance of development of products and their nutrient properties. The primary processing of finger millet grains from the harvesting field covers cleaning, destoning and further grinding into flour for preparation of secondary processed products. The secondary processing such as germination, malting, fermentation, puffing, extrusion cooking, baking are adopted for products preparation. The literature on the processing methods, their optimisation and their impact on processing quality factors such as pH, acidity, role of enzymes, hydrolysis of nutrients, digestibility and fate of non-nutrients have been discussed. The major emphasis is given to processing methods and their suitable modification particularly to finger millet processing.

In: Finger Millet: A Valued Cereal
Editor: K.S. Premavalli
ISBN: 978-1-62081-224-2

Chapter 6

FINGER MILLET STARCH AND ITS PROPERTIES

K. S. Premavalli and C. S. Devaki
Food Preservation Division, Defence Food Research Laboratory, Siddharthanagar, Mysore, Karnataka, India

INTRODUCTION

Starch, the major component of cereals legumes and tubers occur in characteristic granular forms of various shapes and sizes. Starch is a partially crystalline polymer with amorphous, intercrystalline and crystalline regions (Yeol Baik, 1997). Thus starch exhibits two phase transitions when heated with water, one is a melting transition of the crystalline region while the other is glass transition of the amorphous region (Slade and Levine, 1998). The digestion of starch is initiated in the mouth by the action of salivary α amylase and most of the starch is digested in the small intestine by the pancreatic α amylase to release maltose, dextrins and small amount of maltotriose and glucose. Maltotriose inturn by the action of isomaltase form glucose and gets absorbed (Filer 1977). Starch is degraded by various enzymes such as α and β amylases, glucoamylase, pullulanase etc. to oligodextrins and glucose, but depends on the type of starch and particle size. Based on the physical form, some of the starch components escapes intestinal digestion, enters colon and gets fermented (Muir, 1999).

Nutritionally, starches are classified based on their solubility and digestibility. Englyst (1992) classifies starch into readily digestible, slowly digestible and resistance starch fractions. Resistance starch called the functional fibre resists hydrolysis by the amylolytic enzymes (Yue and Waring 1998; Champ 2004) and thus escapes digestion in the small intestine. Further, resistance starch has been classified based on their accessibility and granularity (Brouns et al, 2002). In general, cereals do not differ widely on their starch content and the pattern of bonding via 1, 4 and 1, 6 linkages, but may differ in their characteristics. Normally, cereals contain about 75 to 80% starch with about 20% amylose and 80% amylopectin in their composition. However, the characteristics and functional properties inturn the potential health benefits show their specificities.

1. Finger Millet Starch

Finger millet like any other cereal has starch as the major component of dietary carbohydrates and is present at a level of 60 to 75% based on the varieties. Subramanyan et al (1955) have reported 66.2% starch in H22 variety, while Pore and Magar (1979) have reported 70.2 and 69% in A16 and B11 varieties respectively. The total starch content is shown to be 60-65% in finger millet (Wankhede, 1979 and Moruzzi et al, 1991). Premavalli et al (2004) have studied 10 finger millet varieties for their characteristics. The study revealed that the starch content in 3 hilly varieties were 69.3 to 80.1% with an average of 74.83 ± 5.4, while in 7 base varieties, it ranged from 67.9 to 80.5% with an average of 74.5 ± 4.29. Thus the hilly and base varieties did not vary in average starch content, however, the starch content was dependent on the varieties and white variety Indaf 11 had the lowest starch content of 67.9%.

The characteristic nature of native and enzymatically hydrolysed ragi starch has been studied by Mohan et al (2005). They explained that the molecular weight and degree of crystallanity of the starch residues from the enzymatic digests of ragi starch were significantly higher than rice starch. The SEM examination of residues showed a bigger chunks structure. The type of milling has a role on the starch particles and Katti et al (2008) on SEM analysis have found that flours from plate, emery and roller mills contained particles with a large number of starch granules exposed, whereas those of pin and hammer mills comprised intact starch granules with particles having sharp cut edges. Adebowale et al (2005) have reported irregular, polygonal shaped granules with less than 1% cavity both in native and hydrothermal treated starch. The crystalline properties of ragi starch isolated from 6 varieties were analysed by X-ray diffraction and found that the size of the crystalline regions was related to the sharpness of the profiles. The peaks were sharp containing large crystallites, but the extent of crystallanity and size varied with the varieties (Samantaray and Samantaray, 1997). The X-ray diffraction patterns of finger millet flours by different type of milling showed that the crystallanity as well as the crystal size of the starch in flours decreased with the increase in damaged starch content. The hydrothermal treatment of finger millet starch with moisture of 25-30% at 50°C indicated Type B pattern which is characteristic of normal cereal starches. While heat-moisture treatment with 20% moisture and 100°C resulted in Type C pattern which is characteristic of cross linked or legume starches (Adebowale et al, 2005).

The varieties and the processing methods affect the physical state of the grains inturn starch particles. Roopa (2006) has analysed the isolated starch from 10 finger millet varieties using image analyser. Mean particle length ranged from 7.6 to 9.5 pixels and MRI variety had the highest. Skewness, the degree of deviation from symmetry showed higher proportion of smaller particles. While, Kurtosis, the relative flatness of a particle distribution curve was found to be higher for native starch as compared to puffed millet starch which may be due to the gelatinization of starch during puffing. Roopa et al (1998) have reported the complete disintegration of starch granules in dumpling. Roopa (2006) has observed 2 to 4% reduction in starch during puffing which may be due to the gelatinization. The decrease in starch content during fermentation (Antony et al, 1996), germination (Mbete et al, 2000) malting (Nirmala et al 2002) have been reported. The processing of finger millet decreased starch content by 22% and was attributed to hydrolysis of starch into sugars. (Mbithi et al 2002).

2. Finger Millet Starch Fractions

The general, age old concept of starch fractions based on structural linkages is grouped into two components called amylose and amylopectin, normally constitutes to 20 and 80% respectively in cereals. The large variation in amylose content is generally referred as low amylose or high amylose cereal which have an impact on hydrolysis by α amylases. In finger millet, Wankhede (1979) has reported 16% amylose and 84% amylopectin in the variety studied. Premavalli et al (2004) have reported these components in ten varieties of finger millet. Amongst 3 hilly varieties, amylose content was15.9% in VL146 and VL149 varieties and 17.3% in VL204 with 84.1 and 82.7% amylopectin. In 7 base varieties, amylose ranged from 14.2 to 17.9% and amylopectin from 82.1 to 85.8%. But considering the average value of 15.7 and 84.2%, there was no difference between hilly and base varieties.

The starch fractions based on digestibility are nutritionally important because of their impact on the physiological functions. Therefore, based on the functionality, Englyst (1992) classified starch into rapidly digestible, slowly digestible and resistant starch fractions. Rapidly digestible starch is implied to undergo rapid digestion in small intestine and it covers freshly cooked starchy food such as cooked rice. Slowly digestible starch undergo slow digestion or hydrolysis and is common with some of the raw cereals. Resistant starch is resistant to hydrolysis by the amylolytic enzymes and found in potato, raw banana. Arthi et al (2003) have conducted the in vitro starch digestion of cereals and reported the influence of amylose on the starch digestion. They have reported that roti prepared from ragi had highest total starch and rapidly digestible starch. Roopa and Premavalli (2008) have studied the starch fractions in finger millet varieties of 3 hilly region and 7 from base region. The starch fractions of 3 hilly and 7 base varieties of ragi and effect of puffing has been given in Table 6.1 RDS in hilly varieties ranged from 8.4% to 11.2% with an average of 10.0 ±1.4% while in base varieties it ranged from 8.6 to 9.9% with an average of 9.2 ± 0.6. SDS which is measured after 100 min of incubation varied from 32 to 35% and 26 to 30% in hilly and base varieties respectively, while total starch (TS) as measured by enzyme hydrolysis, varied from 44% to 53 % and 39% to 41% in hilly and base varieties respectively. Relatively on average hilly varieties had higher RDS, SDS, and TS as compared to base varieties. Arthi et al, (2003) reported high RDS of 29.5% and low SDS of 3.3% in one variety of ragi. Sagum and Arcot (2000) reported higher SDS and lower RDS in raw rice. However, this is the first report made available on starch fractions of base and hilly varieties of ragi. The effect of processing i.e. puffing on the starch indicated a considerable change in the starch fractions (Table 6.1). In puffed ragi flour, the results indicated that RDS increased by 5 - 18.9% and total available starch decreased drastically which indicated its hydrolysis during puffing. Relatively, the increase in RDS and decrease in SDS was higher in hilly varieties than the base ones, while total starch as slightly lower in hilly varieties. The analysis of variance amongst the raw ragi varieties for starch fractions revealed that the changes were non-significant ($p > 0.05$). However, between the hilly and base varieties, the changes in SDS, TS were significant ($p < 0.05$) while changes in RDS were non-significant.

Table 6.1. Starch Fractions* (%) in Finger Millet Varieties

Varieties	Raw Flour				Puffed Flour			
Hilly Region	RDS	SDS	TS	RS	RDS	SDS	TS	RS
VL146	8.30 ± 0.1	32.28 ± 0.1	44.39 ± 0.3	0.85 ± 0.1	9.67 ± 0.1	21.74 ± 0.2	15.43 ± 0.3	0.69 ± 0.1
VL149	11.12 ± 0.7	35.33 ± 0.1	50.05 ± 0.2	0.87 ± 0.1	12.15± 0.1	24.25 ± 0.3	17.95 ± 0.8	0.64 ± 0.2
VL204	10.41 ± 0.1	33.67 ± 0.1	47.81 ± 0.2	0.86 ± 0.2	12.85 ± 0.1	24.89 ± 0.2	16.11 ± 0..4	0.68 ± 0.1
Ave ± SD	9.95 ±1.45	30.56 ± 1.5	47.81 ± 2.8	0.86 ± 0.1	11.55 ± 1.6	20.39 ± 3.3	11.19 ± 1.3	0.67 ± 0.1
Base Region								
INDAF5	9.60 ± 0.1	27.72 ± 0.1	40.00 ± 0.4	0.85 ± 0.2	10.65 ± 0.4	18.73 ± 0.2	15.45 ± 0.4	0.56 ± 0.1
INDAF8	9.17 ± 0.1	29.43 ± 0.1	41.44 ± 0.1	0.89 ± 0.1	10.09 ± 0.4	19.36 ± 0.3	19.42 ± 0.7	0.71 ± 0.1
INDAF9	8.44 ± 0.2	26.53 ± 0.2	38.66 ± 0.4	1.00 ± 0.1	10.21 ± 0.3	18.96 ± 0.4	15.13 ± 0.2	0.76 ± 0.3
INDAF11	8.62 ± 0.2	30.56 ± 0.5	40.87 ± 0.3	0.82 ± 0.1	9.92 ± 0.1	19.80 ± 0.1	17.28 ± 0.1	0.64 ± 0.2
MR1	9.79 ± 0.2	29.61 ± 0.2	40.74 ± 0.6	0.95 ± 0.3	10.22± 0.1	23.19 ± 0.5	17.41 ± 0.1	0.64 ± 0.2
HR911	9.83 ± 0.1	27.86 ± 0.2	39.77 ± 0.3	0.96 ± 0.1	10.45 ± 0.3	22.13 ± 0.4	19.52 ± 0.3	0.77 ± 0.4
GPU28	9.50 ± 0.1	28.22 ± 0.2	40.70 ± 0.2	0.91 ± 0.3	10.12 ± 0.1	22.02 ± 0.1	19.72 ± 0.7	0.75 ± 0.2
Ave ± SD	9.27 ± 0.5	28.70 ± 1.7	40.31 ± 1.0	0.90 ± 0.2	10.19 ± 0.3	19.59 ± 1.6	17.00 ± 1.9	0.70 ± 0.3

* Mean ± SD.

RDS: Rapidly digestible starch.

SDS: Slowly digestible starch.

TS: Total starch.

RS: Resistance starch.

The effect of processing i.e. germination, autoclaving and fermentation in complementary food with finger millet preparation Mbithi et al (2002) have reported increase of RDS from 10.3 to 18.9% and SDS from 47.2 to 59.8%.

The resistant starch (RS) is the fraction which escapes digestion in the small intestine (Asp 1992) and in large intestine it becomes a substrate for bacterial fermentation resulting in the formation of short chain fatty acids such as acetic, propionic and butyric acids, carbon dioxide, hydrogen and methane. Cummings and Englyst (1991), Macfariene and Cummings (1991) and Scheppach et al (1988) have indicated that fermentation of RS specifically leads to an increase in butyric acid. According to Mangala et al (1997, 1999b), RS in Indaf variety of ragi is present at low levels of 0.008%. Roopa and Premavalli (2008) have reported RS both in hilly and base region varieties. RS ranged from 0.85 to 0.87% and 0.82 to 1% in hilly and base varieties respectively. Relatively hilly varieties had lower RS than the base varieties. The varietal difference in RS was not considerable and on an average 0.88% RS was present.

3. Resistant Starch Fractions

Resistant starch with varied functional properties of low water holding capacity, small particle size, and bland flavour, with zero energy value has several health benefits (Brouns et al 2002). Finger millet contains RS, a functional fiber and the quantity depends on the varieties and is influenced by the processing methods (Roopa et al 2006). Generally, RS is present in foods in three forms namely RS_1, RS_2 and RS_3. (Englyst et al 1992) and RS is mainly present in pulses and tubers. The RS fractions differ in their crystalline structure and Englyst et al (1992) have reported the procedure for fractionating RS. However, no literature is available on the RS fractions in cereals, but, total RS in various foods (Premavalli et al, 2006) and in chemically modified starches (Woo and Seib 2002) have been reported.

The finger millet resistant starch fractions has been reported by Roopa (2006). In finger millet, the fractions RS_1, RS_2 and RS_3 were 3.46, 2.33 and 30.86% respectively, reflecting that non granular, indigestible and crystalline starch RS_3 is on the higher side. During puffing RS_1 and RS_2 decreased while RS_3 increased. RS_2, the ungelatinised form decreased while crystalline starch increased from 30.86 to 41.93%. During puffing, a HTST process, the starch gets gelatinized (Premavalli et al, 2004) and thus the decrease in ungelatinised fraction is explainable. The increase in retrograded starch (RS_3) may provide resistance to amylolytic digestion and impart physiological benefits.

According to SEM analysis, RS_1 showed oval starch granules as well as breakdown of granular structure in native flour, while in case of puffed samples the bigger irregular shaped particles were observed. In RS_2 from both native and puffed samples, a complete disintegration of starch structure and small pores on the surface were seen. In RS_3, the fissures could be seen because the solid regions of the retrograded starch were less accessible and unhydrolysed resulting in pits on the surface. Faki et al (1983) reported the presence of surface pores or fissures on starch granules may be due to the absence of excess enzyme activity. The deep fissures in the starch granules have been reported to be indicative of a strong bond between the starch and the protein matrix (Colonna et al 1980). Thus the morphological features amongst RS fractions differed considerably while RS_2, RS_3 fractions of puffed samples could be seen as a cloudy mass.

The X-ray diffraction of RS_1, RS_2 and RS_3 fractions from native and puffed flour have been studied. RS_1, showed A-type peaks at 15.1°, 17.1° and 18.0° while RS_2 showed B-type pattern and peaks at 12.7°, 19.8° and 26.6°and RS_3 showed V and B type patterns. It has been reported that potato starch can be converted to A or C type using heat/moisture treatments (Eliasson 1996). Mangala et al (1999a) reported B and V type starches in case of total RS in ragi. The differences in crystalline order of starch indicated its susceptibility to pancreatic digestion. In puffed flour, the crystallinity of RS fractions was totally disrupted with the widening of the pattern by typical peaks RS_1, showed A-type, RS_2 showed C type and RS_3 showed characteristic B and V pattern. Thus, the type of crystallinity was comparable to and during puffing the changes did occur to form B, C and V type crystalline structures.

The effect of processing on RS in cereals, legumes, vegetables (Platel and Shurpalekar (1994), in processed foods (Kavitha et al 1998) have been reported out of which finger millet has also been studied. Arthi et al (2003) reported 4.5% RS in ragi roti. RS formed during parboiling, expanding, popping, flaking, roller drying, malting, and autoclaving of ragi has been determined by Mangala et al (1997). RS formation was more pronounced in autoclaving and a fivefold increase was observed during repeated autoclaving i.e. heating and cooling cycles. Sharavathy et al (2001) have reported 11.2% RS in ragi roti. The storage changes reported in processed foods by Namrata et al (2000) showed a significant increase in RS. Mbithi et al (2002) have reported that during processing of complimentary food with 60% finger millet, RS decreased by 50%. Puffing of finger millet decreased RS by 19 to 26% in hilly varieties and 14 to 31% in base varieties (Roopa and Premavalli, 2008). In this study, in order to understand the relationship amongst the fractions, correlation coefficient between RS and RDS, SDS, TS were calculated (Table 6.2). In both raw and puffed samples RS was negatively correlated to RDS and SDS, TS fractions did not correlate well. When hilly and base varieties were considered, only RS and RDS were highly correlated in raw, base varieties. These results gave an idea that RS formation may be more related with RDS fraction. The significance of the changes in RS and other fractions was calculated by students 't' test. The changes were significant at 5% level between RS and SDS, TS in raw samples and with SDS in puffed samples. Further in base varieties, the changes were significant ($p< 0.05$). In order to understand the behavior of starch during processing, control system of processing was adopted using only ragi flour and other ingredients. The processing effect on Starch fractions in model systems has been presented in Table 6.3. RDS level for all the 12 processing methods ranged from 6-23% as against 9.03% for ragi flour. Baking, frying and shallow frying reduced RDS while roasting and pressure cooking enhanced the formation of RDS to about 23% followed by cooking, autoclaving 16%, puffing and malting. Germination slightly increased RDS as compared to raw ragi flour. Repeated autoclaving decreased RDS. However SDS decreased in case of all the processing methods ranging from 5 to 24% as against 28% in raw flour, however the extent of decrease was dependent on the type of processing. Baking, frying, germination and malting brought down the SDS to 4 to 8% while in other methods 14 to 24% decrease was observed.

Table 6.2. Correlation and Test of Significance between Resistant Starch (RS) and Starch Fractions

	Fractions	Raw (Total)	Puffed (Total)	Raw (Hilly)	Puffed (Hilly)	Raw (Base)	Puffed (Base)
Correlation	RDS	-0.105	-0.27	0.56	-0.63	-0.007	-0.370
	SDS	-0.626	-0.39	0.80	-0.60	-0.705	-0.397
	TS	-0.443	0.44	0.68	-0.99	0.597	0.531
't'-test	RDS	1.10	1.57	0.004	0.003	8.92*	1.61
	SDS	8.43*	6.84	0.0003	0.0009	7.33*	5.28*
	TS	3.39*	1.26	0.0006	0.0011	2.00*	1.78

RDS: Rapidly digestible starch.
SDS: Slowly digestible starch.
TS: Total Starch, RS: Resistant Starch.
* Significant at 5% level.

Table 6.3. Processing Effect on Starch Fractions (%) and Resistant Starch (%) in Control System

Processing	RDS (%)	SDS (%)	RS (%)
Cooking	16.01	17.06	0.71
Pressure cooking	23.47	18.53	1.16
Autoclaving	16.29	23.40	0.58
Re-autoclaving	14.26	23.01	0.40
Puffing	10.22	23.19	0.64
Roasting	23.19	14.01	3.08
Baking	1.79	5.43	0.45
Frying	2.77	8.43	0.27
Germination	9.9	5.48	0.58
Malting	10.46	5.01	0.65
Toasting (1)	8.92	22.92	0.36
Toasting (2)	6.02	19.64	0.40

RDS: Rapidly digestible starch.
SDS: Slowly digestible starch.
RS: Resistant starch.

Sagum and Acrot (2000) have reported increased RDS and decreased SDS during boiling and pressure cooking of rice varieties. RS increased only during pressure cooking and roasting while in all other processes RS decreased. Repeated autoclaving further decreased the RS. However, Mangala et al (1999b) showed that repeated autoclaving increased the formation of RS. In addition to that, they have reported that all the processes-popping, roller drying, extrusion, flaking, parboiling, malting increased the RS formation both in rice and ragi flours.

4. Finger Millet Starch Functional Properties

Starch being the major component of finger millet has an important role in the development of products, their processing behavior in terms of gelatinization and their physiology in terms of digestion, absorption and excretion. Starch has numerous useful functional properties for food applications which include thickening, coating, gelling, adhesion, bulking, absorber of water etc. Gelatinisation, retrogradation and pasting are the most important phenomena in bringing the functionalities. Water and fat absorption of cereal flours is an important functional factor during development of processed foods and is also governed by starch of the flour. The varietal changes with reference to water holding capacity (WAC) and fat absorption capacity (FAC) as reported by Roopa and Premavalli (2008) have been given in Table 6.4. The WAC of hilly varieties was higher than that of the base varieties. However, the FAC remained the same. The puffing of ragi increased the WAC by nearly three and half times and FAC by nearly one and half times as compared to the raw samples. Puffing increased gelatinisation of flour by 70-99% (Premavalli, et al, 2004) and probably this may be the reason for easy imbibation of water and resultant increased WAC.

Gelatinisation of starch is an important parameter from digestibility point of view and is a process which recurs during heating. Mangala et al (1997) have reported that powdered ragi samples showed 100% gelatinization at higher temperatures whereas malted and extruded samples showed the same at low temperatures reflecting the advantage of malting and extrusion. Premavalli et al (2004) have reported the percent gelatinization in both native and puffed ragi flours.

Table 6.4. Water absorption capacity (WAC) and Fat absorption capacity (FAC) of Raw and Puffed flours from different varieties (n=10)

Varieties	WAC* (g/g)		FAC* (g/g)	
	Raw	Puffed	Raw	Puffed
Hilly Region				
Vl146	1.59±0.1	4.98±0.1	1.13±0.1	1.85±0.1
VL149	1.48±0.1	4.88±0.3	1.01±0.2	1.62±0.3
Vl204	1.53±0.2	4.91±0.2	1.17±0.2	1.78±0.4
Ave±SD	1.53±0.1	4.92±0.2	1.10±0.2	1.75±0.2
Base Region				
INDAF5	1.38±0.8	4.10±0.1	1.10±0.3	1.25±0.2
INDAF8	1.39±0.6	4.06±0.4	0.97±0.3	1.47±0.2
INDAF9	1.39±0.1	3.49±0.5	1.12±0.1	1.61±0.2
INDAF11	1.28±0.2	4.05±0.1	0.93±0.1	1.64±0.1
MR1	1.45±0.1	4.24±0.3	0.92±0.3	1.32±0.1
HR911	1.27±0.3	4.04±0.1	1.07±0.2	1.25±0.1
GPU28	1.21±0.1	4.15±0.2	1.19±0.2	1.34±0.1
Ave ±SD	1.33±0.1	4.01±0.3	1.04±0.2	1.41±0.2

*Mean ± SD.

WAC: Water absorption capacity, FAC: Fat absorption capacity.

In native flours during milling, 24 to 34% gelatinization was observed while puffing of grains and grinding resulted in 70 to 99% gelatinization. The varietal differences on the pasting behavior of starch has been studied by Roopa (2006) using Brabender viscoamylograph. The sharp increase in viscosity in the range of 72.6 to 81.7°C occurred depending on the varieties and was the gelatinization temperature. But the peak viscosity for hilly varieties was 354 to 391 BU while the base varieties had 350 to 422 BU reflecting that some of the base varieties have greater water binding capacity and granular rigidity. The cold paste viscosity which is achieved after cooling from gelatinization stage was recorded to be 538 to 583 BU in hilly varieties and 450 to 697 BU for base ones. The increased viscosity is as a result of the swollen granules while cooling, brush each other exudating the soluble components causing thickening in suspension. Mohan et al (2005) have reported higher peak viscosity for ragi as compared to rice starch.

Noda et al (2004) in potato starch reported that the pasting characteristics differs depending on the harvesting time. Premavalli et al (2005) have studied the phase transition behavior of finger millet starch at different moisture levels using differential scanning calorimetry. The ratio of 1:3 of starch to moisture was optimum and the span thermogram showed a singlet transition with minimum at 78°C indicating complete gelatinization. The quenching and annealing processes decreased the gelatinization temperature, however, the decrease was more pronounced in case of quenching from 78 to 48°C. The gelatinization temperature 98°C for ragi flour, 111°C for the mixture of starch, protein, fibre were higher than the starch alone reflecting that the additives such as fibre, protein increase the gelatinization temperature of ragi starch. Roopa (2006) has studied the varietal difference of thermal behavior of ragi starch. The endothermic transition of ragi starch varied with the varieties. In hilly varieties, To, Tm and Tc ranged from 57.3 to 62.7°C, 60.7 to 67.1°C and 71.5 to 77.2°C respectively, while in the case of base one, it was 52 to 68.4°C, 55.9 to 76.8°C and 69 to 89.7°C respectively showing a higher temperature for melting of starch granule crystallites in some of them. The span of endotherm was sharp for hilly varieties (14.2 to 15.1°C) while in base varieties it ranged from 10.9 to 28.1°C. Similarly, the enthalpy of gelatinisation, the energy required for the transformation to this state also followed a similar pattern showing relatively lower enthalpy for hilly varieties (19.3 to 20.85 J/g) as compared to base ones (20.1 to 29.6 J/g). Mohan et al (2005) though found almost similar endothermic transition, the enthalpy of one variety of ragi starch was reported to be higher where as Mangala et al (1999b) reported a lower value. Noda et al (2004) have reported the enthalpy of potato starch to differ with in the varieties and ranged from 19 to 20.8 J/g. The onset and peak temperatures were lower than the pasting temperature measured by amylograph. The varieties showing high transition temperature and high enthalpy for gelatinisation denotes a stronger crystalline nature of the starch granule.

The puffing being a HTST process, the gelatinisation occurred during the process and it ranged from 70 to 99% (Premavalli et al 2004), thus leading to absence of gelatinisation curve in all the varieties. However, the cooling curves gave the nature of crystallites. Glass transition (Tg) has been shown to be an important phenomenon, which determines the properties of bio-polymers (Sperling, 1994 and Jagannath, et al 1999). It was observed that the puffing gave broader amorphous melting compared to native starch. In starch from puffed finger millet, amorphous and crystalline temperatures decreased while percent crystallinity increased ranging from 49.02 to 62.85%. The increased crystallinity is explainable because of high temperature short time (HTST) process adopted during puffing, leading to gelatinisation and retrogradation of starch as a result of subsequent cooling. However, the glass transition temperature, (Tg) which refers to the intactness of the molecules, in native ragi it ranged from -3.86 to -2.53°C in hilly and -2.98 to -0.95°C in base varieties where as in puffed ragi it ranged from -7.01 to -4.76°C in hilly and -10.03 to 6.7°C for base varieties. The Tg indicates that at these temperatures, the stability of the components was higher. Tg decreased to a great extent in case of puffed starch samples when compared to native ones. Tg is also related to free volume of the grain. Decrease in Tg indicated that the free volume had increased (Roos 1995). The average percent crystallinity in hilly and base varieties was 48.37 and 46.77% in native and 55.64 and 56.03 % in puffed ragi flours respectively. Anitha and Muralikrishna (2008) have reported 37.3% crystallanity of ragi starch. Lawal (2009) has reported that starch retrogradation reduced reasonly in modified finger millet starch by etherification with propylene oxide.

In vitro starch digestibility (SD) is one of the important criteria for determining the status of starch in cereals and to understand their nutritional properties. The SD of ragi varieties has been reported by Roopa and Premavalli (2008). The SD of raw finger millet flour is around 7 to 9% while in puffed flour four fold increase to 29 to 35% has been reported which may be due to the starch gelatinization during puffing. The increased gelatinization during puffing has been found earlier (Premavalli et al, 2004). The in vitro starch digestibility in cereals and their blends have been studied by Arthi et al

(2003). The results indicated that the amylose influenced the rate of starch digestion and also on the cereal blends.

Puttaraj (1999) studied the rate of starch hydrolysis in cereal based preparations which included ragi as one of the components and starch digestibility varied with the type of food product. The method of processing of foods have a greater role on the starch digestibility because of the differed temperature-moisture-time combinations. Besides these, the ingredients properties also play a role on the digestibility. In order to understand the behavior of starch during processing. Roopa and Premavalli (2008) have studied these changes during various processing methods. SD was found to be about 35-40% during cooking, autoclaving, puffing, followed by 25-30% during pressure cooking, germination and repeated autoclaving. But during, roasting, baking, frying, toasting and malting, SD was found to be 10-12%. Parallely, percent gelatinization was highest during puffing, 97%; autoclaving, 65.4% cooking, 56.3%; repeated autoclaving 65.4% and less than 25% in all other methods of processing. In the processed foods, the SD was high in puffed flour utilized foods showing 32 to 40%, while in foods raw finger millet flour was used, the SD was 9-12%. However, finger millet starch based products had about 24% SD showing the importance of ingredients composition and processing methods.

CONCLUSION

Finger millet starch as any other cereal starch is the major component which has been reported to vary from 60-80% based on the varieties. The molecular weight, degree of crystallanity, gelatinisation behaviour, structural characteristics, etc., are influenced by the processing methods. The starch fractions on structural linkages are similar to other cereals, but the starch fractions based on digestibility differs. The first report on these fractions based on digestibility as rapidly digestible starch, slowly digestible starch and enzyme hydrolysed starch routes from DFRL, Mysore. Finger millet contains about 30% of slowly digestible starch which releases sugar slowly, thereby, makes it suitable for diabetic personnel. Resistant starch, a functional fibre with small particle size, bland flavour and low water holding capacity has several processing as well as health benefits and is present in finger millet. Fractionisation of finger millet resistant in finger millet resistant starch into RS_1, RS_2, RS_3 and their properties report emerged from DFRL, Mysore and no literature is available on the RS fractions in cereals. The functional properties of starch such as gelatinisation, pasting, digestibility characteristics and their impact on physiological actions are influenced by processing methods and the existing literature on this cereal has been discussed. The physiological role on starch fractions of finger millet in human system and their benefits require the attention of researchers.

In: Finger Millet: A Valued Cereal
Editor: K.S. Premavalli ISBN: 978-1-62081-224-2

Chapter 7

Finger Millet Flavour Volatiles

K. S. Premavalli
Head Food Preservation Division, Defence Food Research Laboratory, Siddharthanagar, Mysore, Karnataka, India

Introduction

Finger millet is one of the food grains receiving the attention of agriculturists and in India more than 18 varieties have been cultivated which shows the difference in their physico-chemical characteristics. Amongst the characteristics, colour is one sensory attribute which has been found varied in varieties and the pigmentation is mainly due to the tannins. Finger millet is perceived with mild sweet taste. The sensory attributes refers to colour, aroma, taste, texture, flavor and acceptance which aids in judging the overall quality of food/food products. Generally the flavor is referred to a combination of taste and aroma and is a mixture of volatile and non volatile constituents. The flavouring compounds in foods exert a marked influence on consumer acceptance, consequently on commercial value of products. First and foremost property employed in quality evaluation of the food is flavor for which an immediate response to volatile compounds contributing to smell or aroma is recognized. Thus, the flavor volatiles play a major role in province of acceptability of the processed foods.

Finger millet is generally utilized after milling into flour and further processing such as fermentation, germination, malting, puffing, roasting etc. result in desirable flavours. The flavouring compounds formed have an impact on the products flavor and acceptability. The flavor volatiles of cereals in general and finger millet in particular have been noticeably not received adequate attention. Houghen et al (1971) reported on the head space volatile composition of number of cereal grains i.e. corn, oats, rye, etc. and identified simple alcohols and carboxyls to the tune of 39 compounds. A study on staple grain rice by Bullard and Holguin (1996) reported 73 volatile compounds in unprocessed rice while Donald et al (1978) have identified 112 compounds in wild rice grain. The stale flavor of stored rice has been attributed to higher levels of hexanal and heptanal (Chikubu, 1970) and volatiles of cooked rice has also been studied. Wheat, the next major cereal has been studied by many workers (Houghen et al, 1971; McWilliams and Mackery 1969;). The volatile flavor components of a lightly milled, whole grain, soft wheat were isolated and identified 11 compounds in the wheat flavour essence by McWilliams and Mackey (1969). A review by Maga (1978) has summarized volatile compounds identified in corn, rye, wheat, triticale, barley and rice and suggested potential areas of further investigation. The literature clearly brings out the fact that the flavouring compound of finger millet has not been understood yet and needs the attention of researchers.

The product manufacture calls for various processing methods which will be responsible for imparting the flavor into the product. Many workers have attempted to study the effect of processing of cereals specifically roasting. An early study by Hridlicka and Janicek (1964) reported the presence of mainly carbonyls in toasted oat flakes. The volatiles associated with roasting of barley have been reported by Wang et al (1968) and Collins (1971). Wang et al (1968) trapped volatiles from ground barley along with those formed during the roasting of barley. The same carbonyl compounds were identified in both the systems. They postulated that the compounds resulted from carbohydrate and amino acid interaction during roasting. Removal of the volatile carbonyls from roasted barley resulted in a product that did not have characteristics barley aroma and thus the authors conclude that carbonyls play an important role in the flavor of roasted barley. In later study Wang et al (1969) investigated the volatile basic fraction of roasted barley and reported the presence of five pyrazine compounds, pyridine and ammonia. Further, Wang et al (1970) identified two sulphur compounds, seven short chain volatile fatty acids, seven alcohols, and three lactones from roasted barley. While, Collins (1971) has identified eight aldehydes, four ketones, and two alcohols and nine pyrazines along with pyridine in the basic fraction of roasted barley. Another group of Japanese workers have investigated the volatiles associated with neutral and acidic fraction of roasted barley and identified as carbonyls.

(Shimizu et al, 1970a,b). Reviews by Maga (1976, 1982) have reflected the intensive investigations on wheat based products mainly bread and other baked products. Kannur et al (1974) have separated the flavouring compounds of chapathies and reported that carbonyls are formed as the major flavoring compounds. Arya et al (1976) have reported that amino acids and sugars are precursors of carbonyls in chapathies. Premavalli and Arya (1983) studied the volatiles formed during roasting of semolina and showed that pyrazines increase during roasting. Yajima et al (1979) have reported highest acidic fraction in cooked rice than basic or oxygenated compounds. The volatile profile of raw and processed rice and corn has been studied using GC-MS and identified aldehydes, alcohols, and ketones (ACS Abstract, 1977). Lam and Proctor (2003) have opined that milled rice surface lipids are important for rice flavor quality. Vasundhara and Parihar (1979) have identified four pyrazines in roasted finger millet. Lasekan and Lasekan (2000) have reported on the volatile flavor compounds in Nigerian millet based beverage called Kununzaki. Twenty eight volatile compounds identified were pyrazines, ketones, alcohols, aldehydes, hydrocarbons and pyrroline. Aldehydes and pyrazines contributed a much higher percentage to the beverage volatiles than other compounds. Flavour volatiles of Uji, an east African sour porridge of both unfermented and fermented products have been reported by Onyango et al (2004a). Flour blends of finger millet with corn, sorghum and cassava were used for fermentation. Pentanal, hexanal, hexadecanoic, 9,12 octadecadienoic, oleic acids were found to be native to the flour blends. Thus the volatile compounds in cereals are quite complex and the complexity increases during processing and product manufacture.

The existing literature reveals that the flavor volatile of cereal grains and their processed products in general constitutes to carbonyls, alcohols, furans, ketones, esters etc. However, no literature is available on the flavor volatiles of millets in general and finger millet in particular except a report of pyrazine compounds present in roasted ragi flour (Vasundhara and Parihar, 1979). In the early years, processing methods such as cooking, germination and malting have been followed during finger millet processing. But with the advancement of varieties, utilization and interest in developing and preservation of new products, the other methods i.e. baking, puffing, roasting have received the attention of researchers. Inspite of varied sphere of studies, flavor of finger millet and their fate during processing has still remained open for research.

Table 7.1. Volatile Profile (%) of Raw Finger Millet Varieties

Varieties	Total Volatiles	Ether extract	Chloroform extract	Basic Fraction	Neutral fraction	Phenolic fraction	Free acid fraction
2300ft asl							
MR1	1.01	0.72 (84)	0.31 (11.6)	34.9	44.2	16.0	4.9
HR911	0.91	0.71 (77.9)	0.20 (22.1)	36.5	43.0	12.5	5.9
Indaf 5	0.61	0.41 (68)	0.20 (32)	67.3	31.9	0.6	0.39
Indaf 8	0.38	0.34 (91)	0.04 (9)	65.2	31.0	1.2	0.6
Indaf 9	0.32	0.31 (96)	0.01 (4)	42.5	49.0	2.9	2.6
Indaf 11	0.53	0.33 (62)	0.20 (38)	42.6	36.9	11.4	5.7
GPU 28	0.29	0.28 (98)	0.01 (2)	42.1	28.3	17.6	9.4
5500 ft asl							
VL 146	0.33	0.28 (87)	0.05 (13)	35.0	51.0	1.0	0.3
VL-149	0.29	0.25 (87)	0.04 (13)	46.9	34.4	6.2	3.1
VL204	0.30	0.26 (90)	0.04 (10)	55.0	58.0	1.8	1.4

Figures in parentheses gives the percentage.

Table 7.2. Carbonyls and Phenols Profile in Raw Finger Millet Varieties

Varietie	Carbonyls (ppm)	Phenols (ppm)
2300 ft asl		
MR1	23.5	14.7
HR911	9.5	11.3
Indaf 5	3.5	10.0
Indaf8	2.5	12.2
Indaf9	3.8	10.8
Indaf11	3.5	10.0
GPU28	5.7	8.8
5500 ft asl		
VL146	2.0	8.6
VL149	2.5	8.5
VL204	5.5.	6.0

Table 7.3. Carbonyls and Phenols Compounds of Raw Finger Millet MRI Variety

Carbonyl Compounds

Peak No.	Retention Time (Minutes)	Concentration (μg %)	Identification
1	0.94	118.6	Not identified
2	2.54	10.8	Acetaldehyde
3	3.08	12.8	Propanal
Phenolic Compounds*			
1	2.52	72.4	Not identified
2	3.11	24.0	
3	3.60	47.0	
4	4.49	295.4	
5	5.89	45.3	
6	6.21	70.6	
7	7.98	72.4	
8	9.35	94.7	
9	11.10	293.6	
10	14.30	72.8	

* μg % as 2-methoxy 4-methyl phenol.

1. Volatiles of Raw Finger Millet

The volatiles of raw finger millet of different varieties grown at different altitudes have been reported (Premavalli and Shobha, 2006). The total volatiles of seven varieties grown in base region (2300 ft above sea level) ranged from 0.178 to 0.73% with highest in MR-1 while HR-911 and Indaf-5 were better and others had very low volatile content (Table 7.1). White variety Indaf-11 showed low volatile content. Further, the extractives in ether, the nonpolar solvent and chloroform, the polar solvent showed higher volatiles in ether extractives irrespective of the volatile content. However, the quantum varied with the varieties and followed the same pattern as total volatiles. In MR-1, HR-911, Indaf-11, Indaf -8, Indaf-9 and GPU-28 the ether extract, contributed to 91-98.8%, while Indaf-5 relatively showed lower content of 60.2%. The chloroform extractives were found to follow the reverse pattern. Fractionation of ether extractives (Cronin 1982) in to four fractions basic fraction 17.5 to 52.8%) neutral fraction (29.6 to 35.5%) and phenolic fraction (1 to 30.3%)

being higher than free acids (0.35 to 18.7%) showed difference amongst the varieties. In the case of varieties grown at altitude (5500 ft above sea level) similar pattern was observed, but varied with the three varieties studied. However, altitude grown varieties had lower levels of volatiles and probably due to lower temperatures during cultivation. Thus, the first report on raw finger millet flavor infers that volatile fractions vary with the varieties and the place of cultivation.

Carbonyls concentration in ether extractives of raw finger millet varied with the varieties. The highest concentration of 23.5 ppm was recorded in MRI followed by HR 911, GPU 28 and Indaf varieties. The altitude varieties had shown relatively lower levels of carbonyls (Table 7.2) similar pattern was found in phenol content. The carbonyls as DNPH derivatives separation by gas chromatography using SE-30 column at 250°C with the FID detector and carrier gas flow of 40 ml per minute identified acetaldehyde and propanal in raw finger millet. The separation of phenolic compounds on SE 30 column at 75°C resulted in 10 peaks with the peaks at 4.49 and 11.1 minutes as the major ones. But, the peaks could not be identified with the standards on hand. The first study reported on the flavor volatiles of raw finger millet leads the information that carbonyls and phenols are the volatile compounds present.

2. Effect of Processing on Volatiles of Finger Millet

In general, processing of foods leads to better digestibility, taste modification, flavor formation, improved texture and impart safety by eliminating microbes. Finger millet basically undergo primary processing of milling into flour to utilize for processing of roti, dumpling. Germination of finger millet is followed for development of weaning foods. Malting is practical for the preparation of 'Ragimalt' to use as a beverage with milk addition. The awareness on the strength of finger millet has led to the conceptulisation of various products demanding different processing methods such as puffing, baking, roasting and frying which are high temperature adoptable methods which surely aids in the formation of flavours. The puffing or popping leads to several changes in characteristics such as gelatinization of starch (Premavalli et al, 2004) intense flavor formation, reduction in antinutrient factors (Wadikar et al, 2006) as well as impart texture improvement and advantage of cold water reconstitution in the products.

Table 7.4. Volatile Profile of Puffed Finger Millet Varieties

Varieties	Total Volatiles	Ether extract	Chloroform extract	Basic Fraction	Neutral fraction	Phenolic fraction	Free acid fraction
2300ft asl							
MR1	1.01	0.72 (84)	0.31 (11.6)	34.9	44.2	16.0	4.9
HR911	0.91	0.71 (77.9)	0.20 (22.1)	36.5	43.0	12.5	5.9
Indaf 5	0.61	0.41 (68)	0.20 (32)	67.3	31.9	0.6	0.39
Indaf 8	0.38	0.34 (91)	0.04 (9)	65.2	31.0	1.2	0.6
Indaf 9	0.32	0.31 (96)	0.01 (4)	42.5	49.0	2.9	2.6
Indaf 11	0.53	0.33 (62)	0.20 (38)	42.6	36.9	11.4	5.7
GPU 28	0.29	0.28 (98)	0.01 (2)	42.1	28.3	17.6	9.4
5500 ft asl							
VL 146	0.33	0.28 (87)	0.05 (13)	35.0	51.0	1.0	0.3
VL-149	0.29	0.25 (87)	0.04 (13)	46.9	34.4	6.2	3.1
VL204	0.30	0.26 (90)	0.04 (10)	55.0	58.0	1.8	1.4

Values in parentheses gives percentage.

Table 7.5. Carbonyls and Phenols Profile in Puffed Finger Millet Varieties

Varieties	Carbonyls (ppm)	Phenols (ppm)
2300 ft asl MR1	35.0	17.5
HR911	13.5	18.7
Indaf 5	5.5	14.7
Indaf 8	3.0	22.2
Indaf 9	10.0	15.4
Indaf 11	4.0	17.0
GPU -28	11.5	17.5
5500ft asl VL146	2.5	13.9
VL149	3.0	14.0
VL204	11.4	14.0

Table 7.6. Carbonyls; Phenols and Pyrazine Compounds in Puffed Finger Millet MR1 Variety

Carbonyl Compounds

Peak No.	Retention Time (Minutes)	Concentration (µg %)	Identification
1	0.95	33.9	Not identified
2	1.6	87.1	Not identified
3	2.54	34.2	Acetaldehyde
4	3.1	48.2	Propanal
5	4.4	29.2	Isovaleraldehyde
6	6.9	43.1	2,4-hexadienal
7	8.0	87.4	Not identified

Phenolic Compounds*

Peak No.	Retention Time (Minutes)	Concentration (µg %)	Identification
1	1.4	62.2	Not identified
2	1.7	69.9	
3	2.1	32.7	
4	2.5	168.1	
5	3.2	90.3	
6	3.7	96.2	
7	4.4	370.0	
8	5.6	42.1	
9	6.1	70.7	
10	7.6	113.3	
11	9.3	28.8	
12	10.2	361.2	
13	11.1	119.6	

Pyrazines**

Peak No.	Retention Time (Minutes)	Concentration (µg %)	Identification
1	0.9	36.1	Not identified
2	1.3	13.6	2-methylpyrazine
3	1.7	27.8	Not identified
4	2.0	30.4	Not identified
5	2.6	11.8	2,5-dimethylpyrazine
6	3.0	10.2	Not identified
7	3.3	33.7	2,3,5-trimethylpyrazine
8	4.0	37.4	Not identified
9	5.1	44.4	Not identified
10	6.0	14.7	Not identified
11	6.6	22.7	Not identified

*µg% as 2-methoxy 4-methylphenol.

** µg% as 2, 5-dimethylpyrazine.

Hence, the studies on flavor volatiles of puffed fingermillet revealed the increased formation of volatiles by nearly one and half times during puffing as compared to raw grain and the increase is irrespective of the varieties (Table 7.4).

Relatively, the hilly grown varieties have shown lower levels of volatiles. The ether extractives forms the major portion and thus may be responsible for flavours. The ether extractives fractionation reflected the pattern of highest of basic fraction followed by neutral, phenolic and free acid fractions, but slight changed sequence was observed with some varieties. The estimation of the volatile compounds by spectrophotometry in terms of carbonyls and phenols revealed the fact that formation of these compounds is based on varieties. The carbonyls ranged from 2.5 to 35 ppm while phenols ranged from 13. 9 to 22.2 ppm (Table 7.5) which has confirmed the results of increased levels on puffing of finger millet. Surprisingly, the phenols concentration do not tally exactly with phenolic fraction in quantum levels which reflect on the structural compact of the grain with reference to variety and cultivation practices also have a role in the formation of volatiles.

The separation of volatile compounds by gas chromatography using SE-30 column further opened up the picture on volatiles in finger millet.Though seven carbonyl compounds were found, four could be identified as acetaldehyde, propanal, isovaleraldehyde and 2,4-hexadienal, out of 11 pyrazines formed, 3 could be identified as 2-methyl pyrazine, 2,5-dimethyl pyrazine and 2,3,5-trimethyl pyrazine (Table 7.6). However, the phenolic compounds remained unidentified. Thus, the puffing of finger millet has formed carbonyls, phenols and pyrazines. The study also has revealed that pyrazines are formed during processing which was not found in raw finger millet.

3. Mechanism of Formation of Flavour Volatiles in Finger Millet

The mechanism of formation of flavour volatiles can be well understood with the suitable model system. Since major component of finger millet is starch with the others being protein and sugar, a model system with finger millet MR1 variety starch has been formulated along with glucose and amino acids which may be the precursors for volatile formation. Volatile profile in model system (Table 7.7) confirms that ether extractives are the major fractions contributing to about 99% with 1% as chloroform extractives. Glutamic acid system showed the highest as 99.3%. In the model system, carbonyls formation is seen increasing with the amino acid, glucose and their combination. The highest levels of carbonyl formation was found in glutamic acid system as 1.5ppm with lowest with cystine amino acid, 0.55 ppm (Table 7.8). Thus the carbonyl profile revealed that formation of carbonyls varied depending on the combination of amino acid with starch and glucose. The highest in case of glutamic acid system followed by leucine also gives the idea of mechanism of formation of strong desirable flavor during puffing of finger millet. This may be due to the higher content of glutamic acid and leucine found in raw finger millet (Hulse et al 1980). The formation of carbonyls as a result of thermal degradation of carbohydrates has been reviewed by Fagerson (1969) and thus only in control system of starch, glucose, amino, acid carbonyls are formed to a lesser extent.

Table 7.7. Volatile Profile in model System of Millet Starch + amino acids and Glucose

System	Ether extract	Chloroform extract
Starch (s)	0.203 (98.8)	0.005
S+amino acid	0.285 (98.9)	0.005
S+Glucose(G)	0.37 (98.3)	0.004
S+G+ Alanine	0.37 (98.3)	0.004
S+G+Leucine	0.301 (99)	0.003
S+G+Glutamic acid	0.41 (99.3)	0.003
S+G+Arginine	0.205 (98.3)	0.003
S+G+Lysine	0.26 (99.2)	0.002
S+G+Cysteine	0.21 (98.6)	0.003

Values in parentheses gives percentage.

Table 7.8. Carbonyls and Pyrazine Profile in Model System

System	Carbonyls* mg%	Pyrazines ** mg%
Starch (s)	0.1	-
S+amino acid	0.3	-
S+Glucose(G)	0.4	-
S+G+ Alanine	0.75	1.31
S+G+Leucine	1.25	2.41
S+G+Glutamic acid	1.5	4.11
S+G+Arginine	0.75	1.46
S+G+Lysine	1.1	2.18
S+G+Cysteine	0.55	2.16

* Expressed as Valeraldehyde.

** Expressed as 2,5-dimethyl pyrazine.

Table 7.9. Identification of Pyrazines (mg %) in Model System

Starch+ Glucose+ Amino acid	Total Pyrazines mg%	Pyrazine	2 methyl Pyrazine	2.5 -dimethyl Pyrazine	2 ethyl 5-methyl Pyrazine	Un-identified
Alanine	1.31	-	-	-	-	1.31
Leucine	2.41	0.39	-	0.048	-	1.97
Glutamic acid	4.11	0.15	0.068	0.14	0.21	3.54
Arginine	1.46	-	-	0.046	-	1.41
Lysine	2.18	0.54	-	-	-	1.64
Cysteiene	2.96	0.37	-	0.35	0.15	2.09

The mechanism of formation of flavour components may be through streckar's degradation as found in non-enzymatic browning. However, the dependence of type of aldehyde formed is based on the amino acids, but the quantity and other carbonyls formation needs the further investigation. Arya et al (1976) have studied the mechanism of formation of carbonyls in chapathies and reported that it is through streckar's degradation. Formation of carbonyls through sugar-amino acid interaction during bread baking has also been reported (Rooney et al, 1967).

Pyrazines are the compounds formed during processing of products through thermal degradation and encompass in non-enzymatic reaction pathways. Amino acids and sugars are the precursors of pyrazine formation (Wang and Odell 1973; Shigematsu et al, 1972; Shibamato and Bernhard, 1977). In the model system of finger millet starch, glucose and amino acids, pyrazines are formed showing highest in glutamic acid system, 4.11 ppm followed by leucine, 2.41ppm cysteine, 2.16ppm and others (Table 7.8). The concentration of pyrazine is again may be related to the strong desirable flavor in puffed finger millet because of relatively high glutamic acid, leucine present in finger millet. The separation by gas chromatography and identification with retention time of standard pyrazines has shown the diversity with the amino acids present (Table 7.9). In glutamic acid system, four pyrazines i.e. pyrazine, 2-methyl pyrazine, 2,5-methyl pyrazine and 2,-ethyl, 5-methyl pyrazine were identified. While in cystiene system 3 compounds i.e. pyrazine, 2,5-dimethyl pyrazine and 2-ethyl, 5-methyl pyrazines were identified. In leucine system, 2 compounds where as only one compound in lysine, arginine system, besides no compound identified in alanine system. The maximum number of compounds was found in glutamic acid system. Some reports are also available on pyrazine generation to be high in glutamine containing systems. (Huang et al, 1995; Yoo and HO, 1997; Chun and HO, 1999). Thus the studies reveal the formation of pyrazine during puffing but not found in raw finger millet.

4. Effect of Processing on Pyrazine Formation

The volatile compounds in cereals are quiet complex and the complexity increases during processing and the product manufacture, pyrazines were found to be the major volatile component in processed cereals.(Withycombe et al, 1976; Lee, 1977; Harding et al, 1978; Tsugita et al, 1978;). The effect of various processing methods on pyrazine formation has been studied in control process system by subjecting raw ragi flour to different processes viz. cooking, autoclaving, germination, roasting, baking, frying, and puffing. Large quantity of sample (one kg) was used for extraction of pyrazines by Likens-Nickerson's apparatus so that is aides in identification of compounds. The pyrazine formation was highest with puffing followed by frying, baking, roasting, germination with the least concentration during autoclaving and cooking (Table 7.10).The results indicate the formation is mainly processing temperature dependent. However, the germination at room condition forming pyrazines needs a further investigation to have a deeper understanding of mechanism. The pyrazine compounds separated by GC showed 8 to 11 pyrazine compounds separated except during cooking.

Table 7.10. Effect of processing on formation of Pyrazines inModel System

Processing method	Processing conditions	Total Pyrazine	Pyrazine	2.methyl Pyrazine	2,5- dimethyl Pyrazine	2 ethyl 5-methyl Pyrazine	Un-identified
Cooking	100°C, 5 min	0.061	-	-	-	-	0.061
Autoclaving	121°C, 10 min	0.81	-	-	-	-	0.81
Frying	160°C, 3 min	5.71	-	-	-	0.016	5.69
Roasting	180°C, 5 min	3.49	0.60	0.84	0.77	1.15	0.13
Baking	180°C, 20 min	4.73	-	-	-	-	4.73
Puffing	220°C, 1 min	6.29	0.81	0.33	0.62	0.83	3.7
Germination	18-30°C, 2 days	0.34	-	-	-	-	0.34

It is observed that, during puffing and roasting, amongst the separated peaks, four could be identified as pyrazine, 2-methyl pyrazine, 2,5 dimethyl pyrazine and 2-ethyl, 5-methyl pyrazine while during frying only 2-thyl, 5-methyl pyrazine could be identified. In all these systems temperatures were as high as 160 to 220°C were adopted and pyrazines are known to form through Maillard reaction at high temperatures. However, though baking is carried out at about 180°C, the higher Rt compounds were formed and were not identified. During cooking, autoclaving and germination, though the temperatures ranged from 30 to 121°C, pyrazines were formed but, relatively in low concentration and it remained unidentified. Vasundhara and parihar (1979) have also reported 4 plyrazineas in roasted ragi flavor. Regarding the ragi cereal this type of detailed study has been reported for the first time.

CONCLUSION

Finger millet with mild sweet taste possesses a dark colour due to tannins and do not show a distinct flavour. Flavour being a combination of taste and aroma is a mixture of non-volatile and volatile constituents and the information on the volatile compounds being unavailable in literature needs the attention of researchers. In this Chapter, the first study reported on the flavour volatiles by DFRL, Mysore reveals that volatile fractions of raw finger millet vary with the varieties and the place of cultivation and carbonyls, phenols are the volatile compounds present in finger millet. The puffing of finger millet increased the volatiles by nearly one half times irrespective of the varieties and compounds identified comprise of carbonyls namely acetaldehyde, propanal, isovaleraldehyde, 2,4 hexadienal, pyrazines namely 2-methyl pyrazine, 2,5 dimethyl pyrazine, 2,3,5 trimethyl pyrazine, phenolic compounds remained unidentified and pyrazine formation during puffing of finger millet is reported now. The mechanism of formation of volatile compounds of finger millet is through sugar-amino acid interaction and maximum pyrazines being formed in glutamic acid system. Finger millet being rich in glutamic acid, thermal intensive processes leads to the formation of maximum volatiles during processing.

The effect of various processing methods on pyrazines formation revealed that the pyrazines were highest with puffing followed by frying, baking, roasting, germination followed by least concentration during autoclaving and cooking and two more pyrazines were identified at high temperature of processing. The limited study on the flavour volatiles of finger millet gives the scope for future research.

In: Finger Millet: A Valued Cereal
Editor: K.S. Premavalli
ISBN: 978-1-62081-224-2

Chapter 8

Finger Millet Based Processed Foods

K. S. Premavalli and Y. S. Satyanarayana Swamy
Food Preservation Division, Defence Food Research Laboratory,
Siddharthanagar, Mysore, Karnataka, India

Finger millet is known for its excellent storage properties and valued cereal nutritionally, physiologically, medicinally and functionally which reflects the structural and physico-chemical characteristics of the grain. Finger millet is a rich source of calcium, iron (Chapter 3), dietary fibre (Chapter 4) and is a good source of slowly digestible starch and resistant starch, a functional fibre (Chapter 6). Over the decades, agricultural researchers have brought out many varieties (Chapter 2), hybrid forms, low tannin grains called white ragi. Therefore, finger millet stands as a competitive cereal for processing of foods. The various processing methods adopted for finger millet (Chapter 5) are at par with the other cereals such as rice, wheat and their products processing. To-day, the market is exposed to variety of processed products from cereals, legumes, seeds and nuts. But in general, the millet processed products and in particular finger millet processed foods are limited or non-existing. The strength of the finger millet can bring out the value added products and the present scenario on the finger millet processed products will be discussed in this chapter.

The primary processed finger millet products utilized over the ages is in the form of raw flour, later puffed flour and ragi malt which provides more a convenience rather than the ready product to the consumer.

Table 8.1. List of Finger Millet Products and Processing Method

Sl.No	Food Product	Product Status	Processing method	Reference
1	Ragi ambali	RTD	Fermentation and Cooking	Rajalakshmi and Geervani, 1990
2	Ajon	RTD	Fermentation and Germination	Patrick and Tom, 1994
3	Kidongo	RTD	Fermentation and Germination	Patrick and Tom, 1994
4	Mangisis	RTD	Malting and Fermentation	Zvauya et al, 1997
5	Tempe	RTR	Fermentation	Mugula and Lyimo, 1999
6	Uji Mix	RTR	Fermentation	Onyango et al, 2000
7	Chhang	RTD	Fermentation	Basappa, 2002
8	Togwa	RTR	Fermentation	Mugula, 2002
9	Kodo Ko Jaanr	RTD	Fermentation	Thapa and Tamang, 2004
10	Weaning food	RTR	Germination	Malleshi and Desikachar, 1982
11	Togwa	RTD	Germination	Mtebe et al, 1993
12	Burfi	RTE	Germination malting concentration	Sangita and Sarita, 2000
13	Kunan Zaki	RTD	Cooking	Lasken and Lasekan, 1999
14	Weaning food	RTC	Precooking and Extrusion	Malleshi et al, 1996
15	Papad	RTF	Cooking and Dehydration	Vidyavati et al, 2004
16	Vermicelli	RTC	Extrusion, dehydration and toasting	Sudha et al, 1998
17	Pasta	RTC	Extrusion	Devaraj et al, 2006
17.a	Millet pasta	RTC	Exrtusion	Roopa, 2006
18	Sweetened millet mix	RTR	Puffing	Premavalli et al, 2003
19	Spiced millet mix	RTR	Puffing	Premavalli et al, 2003
19.a	Millet beverage mix	RTR	Puffing	Premavalli et al, 2003
19.b	Millet functional beverage mix	RTR	Puffing	Premavalli et al, 2003
Sl.No	Food Product	Product Status	Processing method	Reference

Sl.No	Food Product	Product Status	Processing method	Reference
20	Millet laddu	RTE	Roasting	Premavalli et al, 2006
21	Millet sweet cookies mix	RTB	Milling, puffing, mixing	Premavalli et al, 2006
22	Millet functional cookies mix	RTB	Milling, puffing, mixing	Premavalli et al, 2006
23	Millet sweet cookies	RTE	Milling, puffing, mixing	Premavalli et al, 2006
24	Millet functional cookies	RTE	Milling, puffing, mixing	Premavalli et al, 2006
25	Millet Shankarapole mix	RTE	Milling and optimization	Roopa, 2006
26	Millet dosa mix	RTC	Milling and optimization	Premavalli et al, 2006
27	Millet roti mix	RTC	Milling and optimization	Premavalli et al, 2006
28	Millet ada mix	RTC	Milling and optimization	Premavalli et al, 2006
29	Millet chapathi mix	RTB	Milling and optimization	Premavalli et al, 2006
30	Millet pakoda mix	RTF	Milling, optimization and dehydration	Premavalli et al, 2006
31	Millet nippattu mix	RTF	Milling, roasting and dehydration	Premavalli et al, 2006
31.a	Millet jamun mix	RTF	Milling and optimization	Premavalli et al, 2006
32	Millet enteral feed mix	RTR	Milling and optimization	Roopa, 2006
33	Millet halbai mix	RTC	Starch isolation, milling and optimization	Premavalli et al, 2006
34	Millet kheer mix	RTR	Starch isolation, milling and optimization	Premavalli et al, 2006

RTD: Ready to drink; RTR: Ready to reconstitute; RTE: Ready to eat; RTC: Ready to cook; RTF: Ready to fry; RTB: Ready to bake.

The hydrothermal processing has been tried for better properties of the flour (Usha and Malleshi, 2011). Though it improves milling efficiency, it enhances the hardness which will have an impact on the quality of product. Decortication, a primary process though removes the seed coat, the grains will be harder than wheat grits. However, the refined flour obtained after decortication can be used for secondary processed products. But the functional polyphenols and dietary fibre will be lost at the cost of overcoming the limitation of dark colour. The secondary processed products could arise from the processing methods followed in terms of fermentation, germination, malting, puffing, extrusion of the millet. Traditionally, in India, finger millet flour is used for making ragi balls, roti, dosa. Industrially it is used for the malting and brewing processes. Ragi based malted weaning food; porridge and malt are the familiar products with the Indian population (Malleshi and Desikachar, 1982; Rajlakshmi and Geervani, 1990, Sudha et al 1`998). Ragi based products have been developed worldwide namely weaning foods in Kenya (Muroki et al, 1997), fermented products in Tanzania (Mugula and Lyino, 1999) and Indonesia (Saono et al, 1982) fermented gruels in Sweden (Lorri and Svanberg, 1993b) and alcoholic and non-alcoholic beverages in Zimbabwe (Mwesigye and Okurut 1995) and Uganda (Zvauya et al 1997). Secondly, various processed products preparation are being attempted by Home Science personnel by the incorporation of finger millet flour in the processing of several fried, baked foods. But the colour, texture, taste as well as other properties differ very widely thereby decreases the acceptance. Hence simple incorporation of finger millet flour is not the approach to utilize the nutrients as well as to get value added products. But the stability of the products require the studies on the product characteristics during processing and storage which leads to an established shelf life. A combination of technological, nutritional, microbiological study leading to longer shelf life arise from the constant efforts of food scientists and this serve as a base for commercialization and marketing of the product. Firstly, the positive strength of nutrients i.e. calcium and iron and non-nutrient dietary fibre, non-glutenin composition of finger millet should be utilized. Secondly, the understanding and knowledge of non-nutrients role in malabsorption of minerals is necessary to adopt a suitable processing method for maximum reduction of negative effects. Thirdly, the positive strength of slowly digestable starch leads to antidiabetic cereal status to finger millet. Fourthly, the functional fibre and other functional ingredients such as polyphenols, dietary fibre, to be taken into cognition. Fifthly, the possible health benefits of gastrointestinal comfort, fermentation in large intestine to lower chain acids, prevention of colon cancer makes it more valued cereal and the products processed from such a cereal will be a welcome change to the consumer. However, the colour of the products processed with finger millet has to be accepted with reality, otherwise, it becomes a limitation.

The use of whole grain flour, the difference in starch fractions compared to other cereals necessitates a suitable modification in processing methods to achieve the quality product and on these aspects research has been carried out for the product development. Depending on the processing method adopted the products have been grouped and in brief the list is given in Table 8.1.

8.1. Fermented Products

Fermentation is a process in which carbohydrates are broken down into alcohol and acids and is a process used for the preparation of beverages such as wine, beer, cider. Venkatanarayana et al (1979) reported that finger millet could be used as an adjunct along with barley malt in brewing. Rajalakshmi and Geervani (1990) have studied on the tribal foods of South India and reported that ragi ambali, a fermented and cooked product had increased in thiamine and niacin contents by 50 and 37% respectively after processing as compared to 100 and 50% loss in cooked rice. But in vitro protein digestibility (IVPD) decreased by 24% in finger millet ambli as compared to 20% in Bajra gruel and 12% in cooked rice. But lactic fermentation of finger millet in gruel preparation improved IVPD and non tannin varieties were better (Lorri and Svanberg, 1993a). Patrick and Tom (1995) have discussed the production and consumption of alcoholic beverages in Uganda with the main idea of documenting the traditional methods followed which paves the way for better devises of scientific means to improve their quality and optimize their production methods. Zvaiya et al (1997) have reported traditional Zimbabwean beverage with lactic fermentation using six types of finger millet to prepare fermented food tempe. Tanzanian weaning food with finger millet has been developed along with legumes and nuts as other ingredients (Mugula and Lyimo, 1999). The improved IVPD, protein quality has made it to use as weaning food. East African sour porridge called uji is prepared by lactic fermentation and the effect of drying the fermented product uji was studied by Onyango et al (2000). The dry uji mix was accepted well by the panelists. Chhang is a beer like traditional product, a fermented beverage is consumed extensively in the Sub-Himalyan region by Nepalis and Tibetians. Chhang is an acidic-alcoholic beverage with 2.5 to 3% acidity and 15-16% alcohol, prepared by the fermentation of cooked finger millet using microbial cultures (Basappa, 2002) and on fermentation, all B series vitamin increased. Togwa is a Tanzanian fermented weaning food produced using finger millet, besides other millets. Mugula et al (2003) have studied the microbial and fermentation characteristics of Togwa. Thapa and Tamang (2004) reported on the product characteristics of finger millet beverage, Kodo ko jaanr.

Millet Functional Cookies Mix

Millet Chapathi

Figure 8.1. Finger Millet Based Convenience Mixes Products Profile.

8.2. Germinated Products

Germination process is a promising one for improving the digestibility and bioavailability of nutrients. Malleshi and Desikachar (1982) have formulated finger millet and greengram based weaning food with low hot paste viscosity. The shelf life studies of weaning food in flexible pouches by Malleshi et al (1989) have shown 2 months at 37°C and 3 months at 27°C when packed in lower density polyethene pouches. But the shelf life could be increased to 3 and 5 months when stored at 37 and 27°C in laminated pouches. Hemamalini et al (1980) reported that sprouted ragi during germination acts as a better growth promoter. Refined flour from sprouted ragi called vaddragi flour showed best calcium availability due to lower fiber content. The improvement on the weaning food of India was attempted by optimizing the conditions of germination to get the malted flour (Malleshi and Desikachar, 1982). The optimized conditions of germination at 20°C for 4 to 5 days gave malt with maximum enzyme activity. Dehydration at 70°C to 12% moisture resulted in desirable aroma with retention of α amylase activity and minimum loss of available lysine. The high amylase activity in the malted flour caused slurries of low hot paste viscosities. The weaning food was though good in protein quality, but deficient in threonine (Malleshi et al, 1986)and weaning food with 10% skim milk powder was well accepted. Mtebe et al (1993) studied on the characteristics of Togwa prepared in different parts of Tanzania and reported that togwa from finger millet was the most preferable. The germination of grains have improved the flavor, acceptability of togwa and improvement in bioavailability of nutrients. Sangita and Sarita (2000) have reported on the preparation of burfi from malted flour and the products of four genotypes of finger millet showed acceptability range of fair to good.

8.3. Cooked and Extruded Products

Cooking of finger millet flour is a routine practice followed for cooked balls or gruel preparation for immediate consumption. But for commercialization of products, the preservation with adequate shelf life is necessary. Lasekan and Lasekan (1999) have discussed on non-alcoholic beverage called Kunan Zaki which is widely consumed in Nigeria, Chaol and Sudan. Kunun Zaki involve wet milling of finger millet with spices such as ginger, cloves, pepper and after sieving, gelatinized partially followed by sugar addition and bottling. The product has a characteristic sweet taste and roasty potato like flavor. Malleshi et al (1996) have studied the physical and nutritional qualities of extruded weaning food and was an improvement and blending with wheat flour. Medium size particle flour used for dough gave better quality product and the addition of wheat flour improved the product. Besides extrusion and dehydration, toasting of vermicelli for 3 mins at 120°C improved the cooking quality. Devaraju et al (2006) optimized the composition of finger millet composite flour for pasta with 50% finger millet and the hot water mixing gave better quality pasta.

Table 8.2 Nutritional and Functional Strength of Finger Millet Based Convenience Mixes

Sl. No.	Convenience mixes	Fat, %	Protein, %	Dietary fibre, %	Calcium mg%	Iron, mg %	Energy value kcals/100g	Shelf life, months
1	Sweetened millet mix	20.1	7.8	12.0	144	5.5	500	4
2	Spiced millet mix	20.0	11.4	11.0	209	8.0	460	6
3	Millet beverage mix	1.38	9.9	10.8	98	3.6	407	4
4	Millet functional beverage mix	5.67	6.9	10.6	200	5.5	428	6
5	Millet sweet cookies mix	20.46	9.7	7.0	112	4.2	502	12
6	Millet functional cookies mix	25.33	7.51	7.8	126	4.3	567	8
7	Millet shankarapole mix	8.06	8.23	6.0	120	4.0	427	6
8	Millet dosa mix	1.47	7.5	3.8	330	9.5	390	8
9	Millet roti mix	11.28	6.84	13.2	260	10.1	456	1
10	Millet ada mix	2.88	11.41	12.0	302	5.8	414	2

8.4. Puffed Flour Products

Puffing is an HTST process which pops the conditioned grains, resulting in higher desirable flavour formation. Puffing also leads to gelatinization of starch (Premavalli et al, 2004) and improves digestibility. Premavalli et al (2003) have reported finger millet puffed flour based convenience mixes both sweetened and savory products with a shelf life of 6 and 4 months shelf life at ambient conditions packed in flexible laminate pouches.

The products had the advantage of cold water reconstitution prior to consumption and acceptance was excellent because of highly desirable flavour. Vidyavathi et al (2004) substituted black gram dhal flour by finger millet flour by 50 % during the preparation of papad, which needs to be toasted or fried prior to consumption. Finger millet papad showed a higher acceptability and also serve as good source of calcium at 102 mg%. Premavalli et al (2006) have optimized the ingredients during the development of millet laddu based on finger millet and green gram dhal and the product was liked very much. Millet laddu had an established shelf life 4 months at ambient conditions when packed in laminate pouches.

8.5. Baked Products

Baked products utilisation is on the rise probably because of convenience of ready to eat nature, varieties with varied flavours and taste. Millet cookies convenience mix both sweet and functional have been developed (Premavalli et al, 2006) which has a shelf life of 10 months, but required to be baked. Millet sweet cookies and functional cookies were developed with 50 % ragi flour in the composition of ingredients. The product had good stability, acceptance and its beneficial effects on constant consumption has been studied in school children which will be presented in chapter 9. The use of ragi flour leads to grittness in biscuits which is not desirable and this characteristic property is probably due to the seed coat inclusion in the flour as well as starch properties. The research on these aspects has led to the use of combination of raw and puffed flour for cookies preparation. The functional cookies had higher protein as well as Caroteniods. Selvaraj et al (2002) have studied the effect of packaging and storage and claimed that the acceptability was at par with maida based cookies. However, 20% finger millet flour was incorporated which was very low level for its benefits.

8.6. MILLED FLOUR PRODUCTS

Finger millet utilization from times immemorial is in the form of finger millet flour which is obtained after milling. Milled flour is being used at present too for the preparation of roti, dosa, cooked balls etc. and have as fresh product. The global scenario in the present day is the demand for convenience processed foods, preferably ready-to-eat with better shelf life, satisfying taste, ease of portability, with high nutrition quality are increasing because of growing urbanization and increased employment of women in industrial and public sectors. The significant component of the dietary profile of present day consumers is baked foods and will be of much better nutrition with finger millet which has already been reported in this chapter. Another group of products namely, convenience mixes will be handy to civil sector and these being an integral part of ration packs for Defence Services, it will be essential requirement for Defence Sector. These products require some pre-preparation prior to consumption. Therefore, the development of finger millet based convenience mixes has been achieved by optimizing the ingredient composition by the routine method of permutation-combination of ingredients as well as by response surface methodology using statistical design software for some of the products. (Premavalli et al, 2006). Milled finger millet flour was ranging from 30-91% apart from other ingredients. The finger millet (FM) component was quite substantial unlike other products reported with 15-20%. *Millet dosa mix* with about 90% finger millet flour has a shelf life of 6 months and now slowly gaining importance in the commercial market. The mix should be added to adequate water to make a slurry and toasted on iron thawa. *Millet roti mix* with about 60% finger millet flour have a short shelf life of 2 months and the low shelf life is due to other flours used which may undergo oxidation. *Millet ada mix* is a legume based mix which has been accepted very well and is of good taste with 4 months shelf life. *Millet chapathi mix* has proved its efficiency in terms of acceptance in services. *Millet enternal feed mix* serve as a nutritive drink after reconstitution and is a good addition to the hospital diet profile. Another group of products are fried foods which takes the time for formulations to be made ready prior to frying. Therefore, some of the Indian known fried items in the form of convenience mixes with all the ingredients have been developed and established their shelf life. *Millet nippatu mix,* protein rich snack mix with many ingredients along with 55% of finger millet flour require the preparation of stiff dough with water followed by deep fat frying in circular shapes. *Shankarapole mix* also needs dough preparation, deep fat frying to get ready tasty snack and the mix form has a shelf life of 6 months. *Millet pakoda mix* is a spicy product liked by the people which require deep fat frying process prior to consumption and has a shelf life of 6 months at ambient conditions. Jamun as a sweet is liked by all ages and millet jamun mix is a nutritious, energy dense product which require preparation of dough, then small balls to be deep fat fried and kept in sugar syrup. The most interesting point is in jamun mix, the colour of millet is not a limitation because the fried product will be brown in colour. In all other convenience mixes, the natural colour will be little dark because of finger millet grain colour, but, the studies revealed that it is not a great limitation either for acceptance or shelf life. Still for an eye appeal and for light coloured product, white variety flour can be used. The factor which has a greater impact on shelf life is the unsaturated fatty acid composition of the finger millet as well as the other ingredients role. Finger millet has a starch content of about 76% and the isolated starch was used for the development of *Millet Halbai mix* and *Millet Kheer mix*. Halbai, a sweet dish which traditionally when prepared by steeping the grain, grinding, sieving and concentration with jaggery to pliable consistency require about 2 days. The pre mix for Millet halbai takes about 20 to 25 minutes of cooking to get the pliable product and is a great advantage. Millet kheer mix again a sweet dish require hot water reconstitution for 2-3 minutes which gives a tasty kheer drink. But the limitation is relatively the short shelf life of 2 months and use of better barrier properties packaging improves the shelf life.

These convenience mixes had a moisture content 4 to 6% which is also a causative factor for better shelf life. The finger millet products profile along with the processes adopted is given in Figure 8.1. The nutritional strength with energy value of the convenience mixes is presented in Table 8.2 The products are rich in calcium, good source of iron and dietary fibre. Millet nippattu mix and millet ada mix are protein rich besides other positive strength and provide adequate calories in the diet. Therefore, these strengthful products will be no doubt new additions to the present day list of processed products. The health benefits imparted by the grain as antidiabetic, preventive of colon cancer, better gastro intestinal health can be well enjoyed by utilization of the stable finger millet processed foods. The type of the finger millet food products developed in the past two decades gives the view that alcoholic, non alcoholic beverages, gruels, weaning foods are being systematized for manufacturing practices (Table 8.3). But the fact clearly shown is the development of newer products of finger millet such as pasta, papad, vermicelli and a large number of convenience mixes for routine use of the consumer.

Table 8.3. List of Finger Millet Products Type with Country of Origin

Sl.No	Food Product	Type of product	Country of origin	Reference
1	Ragi ambali	Nonalcoholic gruel	India	Rajalakshmi and Geervani, 1990
2	Ajon	Alcoholic beverage	Uganda	Patrick and Tom, 1994
3	Kidongo	Beer	Uganda	Patrick and Tom, 1994
4	Mangisis	Acidic alcoholic beverage	Zimbabwe	Zvauya et al, 1997
5	Tempe	Weaning food	Tanzania	Mugula and Lyimo, 1999
6	Uji Mix	Acidic beverage	East Africa	Onyango et al, 2000
7	Chhang	Acidic alcoholic beverage	India	Basappa, 2002
8	Togwa	Weaning food	Tanzania	Mugula, 2002
9	Kodo Ko Jaanr	Alcoholic beverage	India	Thapa and Tamang, 2004
10	Weaning food	Ready to cook	India	Malleshi and Desikachar, 1982
11	Togwa	Ready to drink sweet gruel	Tanzania	Mtebe et al, 1993
12	Burfi	Ready to eat sweet	India	Sangita and Sarita, 2000
13	Kunan Zaki	Non alcoholic beverage	Nigeria	Lasken and Lasekan, 1999
14	Weaning food	Cooked gruel	India	Malleshi et al, 1996
15	Papad	Cooked and dehydrated dry product	India	Vidyavati et al, 2004
16	Vermicelli	Dry product, Extruded	India	Sudha et al, 1998
17	Pasta	Dry product, extruded	India	Devaraj et al, 2006
17.a	Millet pasta	Dry product, extruded	India	Roopa, 2006
18	Sweetened millet mix	Puffed, cold water reconstitution	India	Premavalli et al, 2003
19	Spiced millet mix	Puffed, cold water reconstitution	India	Premavalli et al, 2003
19.a	Millet beverage mix	Puffed,cold water reconstitution	India	Premavalli et al, 2003
19.b	Millet functional beverage mix	Puffed,warm water reconstitution	India	Premavalli et al, 2003
20	Millet laddu	Sweet product	India	Premavalli et al, 2006
21	Millet sweet cookies mix	Convenience to bake	India	Premavalli et al, 2006
22	Millet functional cookies mix	Convenience to bake	India	Premavalli et al, 2006
23	Millet sweet cookies	Sweet biscuits	India	Premavalli et al, 2006

Table 8.3. (Continued)

Sl.No	Food Product	Type of product	Country of origin	Reference
24	Millet functional cookies	Savory biscuits	India	Premavalli et al, 2006
25	Millet Shankarapole mix	Snack	India	Roopa, 2006
26	Millet dosa mix	Convenience mix	India	Premavalli et al, 2006
27	Millet roti mix	Convenience mix	India	Premavalli et al, 2006
28	Millet ada mix	Convenience mix	India	Premavalli et al, 2006
29	Millet chapathi mix	Convenience mix	India	Premavalli et al, 2006
30	Millet pakoda mix	Convenient to fry	India	Premavalli et al, 2006
31	Millet nippattu mix	Convenient to fry	India	Premavalli et al, 2006
31.a	Millet jamun mix	Convenient to fry	India	Premavalli et al, 2006
32	Millet enteral feed mix	Convenience mix	India	Roopa, 2006
33	Millet halbai mix	Convenience mix	India	Premavalli et al, 2006
34	Millet kheer mix	Convenience mix	India	Premavalli et al, 2006

CONCLUSION

Earlier traditional products were finger millet raw flour and their products such as ragi balls, dosa, roti and non-alcoholic and alcoholic beverages which were treated as a poor man's diet. With the advent of several cultivars and understanding of grains strengthful properties, in the past two decades, the traditional products are entering the systematised manufacturing practices, development of new products for better functionality has been achieved, with the research orientation on the products characteristics. Over two decade's development, nearly 40 processed foods have been documented, out of which India ranks first. Development of newer products such as pasta, vermicelli and quite a large number of convenience mixes specially from DFRL, Mysore will serve handy to civil sector and more essential to Defence sector for overcoming the problems of skeletal health, diabetes, colon cancer and better gastro-intestinal management.

In: Finger Millet: A Valued Cereal
Editor: K.S. Premavalli

ISBN: 978-1-62081-224-2

Chapter 9

Finger Millet and their Product's Evaluation

K. S. Premavalli and C. S. Devaki
Food Preservation Division, Defence Food Research Laboratory,
Siddharthanagar, Mysore, Karnataka, India

Finger millet is the cereal of choice in India and Africa for the nutritional strength, controlled diseased conditions, gastro intestinal comfort for all the ages with weaning food for children, preferred products for middle age and amblic drink for the aged. The products are more palatable and richness of calcium, dietary fibre makes it more strengthful. However, the dark colour of the grain inturn the products have the limitations of accepting nature's reality. According to the regulations and protocols to be followed, the food products evaluation is an integral part of the product development. In general, the products are evaluated by biological assays with animals which comprise of the measurement of biological value (BV), digestibility coefficient (DC), protein efficiency ratio (PER) and net protein utilization (NPU) etc., after administration of the experimental and control diet. The products evaluation through human subjects will be measured in terms of growth in infants and children and metabolic studies in later age in terms of nutrient levels, as well as diseased condition status such as sugar and cholesterol levels. The evaluation of finger millet products so far reported in both the systems will be dealt in this chapter.

9.1. Animal Studies

The animal studies have been reported using albino rats, to start with finger millet flour, then with the protein supplementation through nuts, legumes, milk powder etc., along with finger millet. Further, the effect of extracts or the diet on physiological action has been reported. Niyogi et al (1934) studied finger millet flour in metabolic trials with rats, with 5% protein in the diet (N=0.84%). The control diet combined 5.0% dried egg powder. Average digestibility and biological value of the protein in the two diets were 77.5% and 90.5% respectively. Ramaiah and Satyanarayana (1936) reported that, by rat growth and nitrogen balance studies, the nutritive value of white finger millet grain is superior to brown type. The same authors (1937) reported the biological value of white finger millet protein as 91.5 superior that of brown type. The rat growth trials was conducted with weanling albino rats using poor vegetarian diet containing 78.5% of the ration as finger millet experimental flours (Kurien,1967). Amongst the diets of white meal flour (A), refined flour (B) and composite flour made of refined flour and added husk, composite flour diet (C) produced significantly higher weight gain than refined flour (B) with lower gains with (A). In calcium availability study, though average intakes were same, the retention from diet whole meal flour was 47% as compared to 68% for (B) and 74% for (C).

Patwardhan (1961) in a literature review reported that the PER of finger millet, fed as the whole grain, was not increased by supplementation with chickpeas or dried powdered amaranth and the PERs were, respectively 2.0 and 2.1. An Indian Multi Purpose Food (MPF) fortified with Calcium and vitamins A, D and riboflavin was compared with an American MPF, when added in poor Indian finger millet diets by Kuppuswamy et al. (1957). The addition of both the MPF significantly increased weight gains compared to the control. Shurpalekar et al (1962a) compared fish flour fortified with vitamins and (Skim Milk Powder) SMP, fortified with vitamins A and D as supplements to poor Indian finger millet diets. Diets (B) fish flour and (C) SMP both produced significant increase in weight gain, haemoglobin and red blood cell counts compared to diet (A) but there were no significant differences in liver fat content among diets (A), (B) and (C). Tasker et al (1964) reported the supplementary effect of a protein food based on a 1.5 blend screw-pressed coconut meal, chickpea and groundnut flour fortified with vitamins A and D, thiamine and riboflavin when added to a poor vegetarian finger millet diet. Supplementation increased both protein content of the diet and the rate of weight gain. Leela et al (1965) examined the effect of supplementing market samples of finger millet and a poor vegetarian finger millet diet with L-lysine, DL-threonine and SMP. The amino acid additions raised the total to the equivalent of egg protein. PERs were determined at both 6 and 8 % levels of dietary protein. The results indicated that finger millet protein is adequate in all essential amino acids except lysine and threonine. The PER of finger millet (FM) protein at the 6% protein levels, diet (A-97% FM) increased significantly supplementation with lysine, (B-97% FM) or lysine and threonine (C-97% FM). The mean liver fat content of rats fed with 6% finger millet protein, (A), was significantly higher than that of rats fed 6% milk proteins (H-SMP). Supplementation of the finger millet diet with lysine and threonine, (B) and (C), decreased liver fat content but not significantly. The PER of the poor finger millet diet fed at the level of 6% protein level, (D-62.5% FM)was significantly higher than that of the finger millet diet (A), at the same level. This may be due to the fact that the lysine content of the proteins of the poor finger millet diet was higher at 256mg/g total N than that of the cereal itself (212mg/g total N). Supplementation of the poor finger millet diet with lysine and threonine produced significant increases in PER when fed at both the 6% and 8% protein levels. Hariharan et al (1967) reported experiments with finger millet in poor vegetarian diets similar to those reported in 1965 with sorghum and pearl millet. Supplementation of the poor finger millet diet with Calcium salts and vitamins did not significantly increase growth rates. Supplementation with groundnut flour (unfortified or fortified) to provide 2.5% or 5% extra protein in the diet, significantly increased weight gain. The PER of the finger millet diet increased slightly with added vitamins and minerals. There was no significant difference in the PER between diets containing 2.5% extra protein from fortified groundnut flour or skim milk powder, but the PER from 5% extra skim milk powder protein was significantly higher than 5% protein from groundnut flour. Fortified and unfortified groundnut flour appeared equally effective in supplementing the poor finger millet diet. Daniel et al. (1968a) compared the additions of chickpeas, pigeon peas and SMP to poor Indian finger millet diets. The average weight gain per week were 18.4 and 20.1g for addition of chickpeas, 17.6 and 18.8g for pigeon peas, 18.7 and 22.7 for soy flour and 21.7 and 24.2 for skim milk powder. PERs were 2.5 and 2.7 for chickpeas, 2.4 and 2.5 for pigeon peas, 2.4, 2.8, 2.5, 2.8 for soy flour and 3.1, 3.0, 2.5 and 2.7 for skim milk powder. Kurien et al (1969) examined the supplementary value of Indian multipurpose food based on a 3:1 blend of groundnut and chickpea fortified with Calcium salts, vitamins A and D and riboflavin, to poor finger millet diets fed to rats, under conditions of adequate and inadequate calorie intake. The rats were fed amounts of diets (A) 8g, adequate calorie intake (C) 6g, calorie intake restricted to 75% (E) 4g, calorie intake restricted to 50%, per rat per day over 4 weeks. The quantity of multipurpose food in (B), (D) and (F) was maintained at 0.8g per rat per day by replacing equal amount of the basal diet. At each caloric intake supplementation significantly increased weight gain and improved Feed/Gain ratios, though only at adequate calories was the PER significantly improved. PER declined with reduction in caloric intakes. The results indicate that multipurpose flour exerts a significant supplementary value when the caloric intake is restricted to 75% but only a slight supplementary effect on 50% caloric restriction.

Desai et al (1970) supplemented poor vegetarian finger millet diets with L-lysine HCl, DL-threonine and DL-methionine. The lysine, threonine and methionine were added at levels designed to raise their dietary content to about the same level as in human milk protein. Supplementation with lysine, diet (B) significantly increase PER over the basic poor diet (A). Addition of (E) lysine plus methionine or (F) lysine plus threonine caused a further significant increase, (H) lysine plus methionine plus threonine, produced the highest PER, significantly higher than (E) and (F). Narayanaswamy et al (1970) added yeast grown on petroleum hydrocarbons to poor Indian vegetarian finger millet diets. The added yeast increased lysine and threonine contents and at the 6% level (2.0% extra protein) with vitamins and mineral mix, diet significantly increased PER. Red gram dhal cooked in an autoclave, dried and milled, was incorporated at 8.5, 16.7 or 25% to replace part of the 83.5% cereal in poor rice or ragi diets and provide 1.5, 3.0 or 4.5% extra protein. The diets were given to weanling albino rats for 4 weeks and protein efficiency ratios (PER) were calculated. The rice and ragi basal diets had 4.6 and 3.4 g lysine per 16 g N and each had 3.9 g threonine per 16 g N. The poor rice diet when supplemented with minerals and vitamins supported mean weekly weight gain of 13.4 g; with the 3 levels of red gram that increased to 17.0, 17.9 and 18.2 g, the last 2 not significantly greater than the first increase. The PER was highest with 8.5% red gram, 3.24 against 3.00 for the basal rice diet. The poor ragi diet with minerals and vitamins supported mean weekly gain of only 5.9 g. The increases with red gram were to 14.8, 17.1 and 18.6 g, and the PER rose with the lowest addition of gram from 1.80 to 2.98. Without the vitamins and minerals the addition of gram to rice diet was of little value but the vitamins and minerals had less effect with ragi diet which probably provided enough of them (Daniel et al, 1970). Narayanaswamy et al (1972) added (1) Low cost protein food (wheat flour 70/soy flour 15/ groundnut flour 15) and (2) fortified skim milk powder to a poor finger millet vegetarian diet. Diets (A-78.5%

finger millet) and (B-75.5% finger millet) provided 8.1 and 8 % protein, diets (C-65.5% finger millet) and (E-69.5% FM) provided 1.5 % and diets (D-55.5% FM) and (F-63.5% FM) 3.0% extra protein. There was significant increase in weight gain, and only a slight increase in PER, between diets (A) and (B). Diets (C), (D), (E) and (F) produced highly significant increases in weight gain and PER compared to (A) and (B). PERs were not significantly different between diets (C) and (D), nor between (E) and (F); were PERs of (E) and (F) significantly higher than (C) and (D). The low cost protein food compared well with skim milk powder as a protein supplement in poor finger millet diets. Daniel et al (1974) reported on the nutritional complementarity of finger millet and chickpeas. A blend of finger millet and chickpeas in 80:20 ratio contained about 9.6% protein, and gave a PER of 2.9. Finger millet and chickpeas separately each adjusted to 6.0% protein with maize starch gave PERs of about 1.2. Narayanaswamy et al (1974) studied the supplementary effect of three protein enriched cereal foods (PECF), blends of wheat flour, soybean flour and groundnut flour, on poor vegetarian finger millet diets. Diet (A-75.5% FM) provided 8.3% protein, (B-65.5% FM), (D-65.5% FM) and (F-65.5% FM) provided 1.5% and (C-55.5% FM), (E-55.5% FM) and (G-55.5% FM) provided 3.0% extra protein in the diets. All supplements significantly increased rat weight gains compared to diet (A) and the mean growth rate and PERs from diets (B),(C),(D) and (E) were significantly higher than diets (F) and (G). Venkataraman et al (1977) reported experiments in which the green algae was added at two levels diets based on finger millet fed to male weanling rats for a period of 4 weeks. The supplementation, which increased the protein content of the diet from 6.0 to 8.6%, significantly increased the PER compared to the finger millet diet alone. The biological availability of calcium from six ragi-based diets was studied in comparison with skim milk and calcium lactate diets. Among the five supplemented ragi-based diets, however, diet 5 was found to result in the highest calcium retention levels when fed to rats (Rajammal et al, 1980a). Sprouted ragi was found to be more efficient in promoting growth of rats than either whole ragi or vaddragi flours. Although no significant differences were observed in protein efficiency ratio or calcium retention between unsprouted and sprouted ragi, the overall nutritive value and nutritional quality of sprouted ragi was much higher due to greater amounts of B-vitamins. Calcium availability was significantly higher in vaddragi, a processed ragi, due to its low fiber content. (Hemamalini et al, 1980)

Mitchell (1922) introduced the term 'Net Utilization of Dietary Protein' which is a product of digestibility coefficient and biological value divided by 100. *Net Protein Ratio* introduced by Bender and Doell (1957) is a modification of the PER method. This consists in feeding a group of weanling rats on diet containing 10% test protein and another group receiving non-protein diet for a period of 10 days. The NPR is obtained by adding together the loss in weight in the non-protein group and the gain in weight of the test group and dividing by the quantity of protein (g) consumed by the test group. This procedure allows for maintenance requirements and also permits the evaluation of poor quality proteins which do not promote any growth and whose PER is zero or negative. Bender and Doell (1957) reported that NPR values correlated closely with NPU values for a large number of foods.

The effect of including ragi husk in the diet was studied by Kanchana and Kantha (1983), at 8 per cent level, on growth and body composition of pair-fed albino rats. Even on equal food intake by both control as well as the experimental animals, the incorporation of ragi husk promoted better growth, nitrogen retention and protein efficiency ratio. Protein content of the carcass and liver of rats fed diets containing ragi husk was higher than the control group. The growth response of rats fed ragi husk appeared to increase with the content of nitrogen in the husk.

The nutritional quality of a malted weaning food developed using malted ragi (*Eleusine coracana*) and green gram (*Phaseolus radiatus*) was evaluated by rat feeding trials. The protein score of the weaning food was 70 calculated according to FAO/WHO (1973) pattern. The protein efficiency ratio (PER), net protein utilization (NPU), biological value (BV) and true digestibility (TD) values for the weaning food proteins at 10% level of protein intake were 2.2, 51.6, 73.8 and 82.8, respectively and the relative protein value (RPV), determined at 5, 8 and 11% levels of protein intake was 84. Supplementation of the weaning food with 10% skim milk powder enhanced the PER to 2.7 and NPU to 63.0. The nutritional quality of a roller dried proprietary weaning food was also evaluated along with malted weaning food for comparison and it was observed that the nutritional quality of the two products were comparable (Malleshi et al, 1986).

Slope Ratio Method: In an attempt to arrive at a method which yields linear regression between dose and response, Hegsted and Chang (1965) developed the slope ratio method for bio-assay of proteins. The 'relative growth index' is the slope of regression between dose and response expressed as a percentage of the slope obtained with a protein of maximal nutritive value. Mason et al (1946) over 9 weeks fed nine male and nine female rats, in pairs, diets based on rice to which varying amounts of butter and/or finger millet were added. Analysis of variance showed that both the negative effect of butter and the positive effect of finger millet on rat growth were highly significant but there was no interaction between butter and the millet; finger millet did not overcome the growth reducing effect of the fat.

Serum protein (King, 1951) was not significantly different among diets but diets (B), (C) and (D) all gave significantly higher haemoglobin contents and red blood cell counts. There was no difference in average liver protein (N X 6.25%) content among the four diets, but diets (B), (C) and (D) produced significantly higher liver fat (Tyner et al., 1950). Over all, the replacement of even 25% of rice by finger millet in a poor vegetarian diet appeared to improve the nutritive value of the diet.

Subrahmanyan et al (1954a) compared the supplementary value of (a) cassava flour, (b) sweet potato flour, (c) a mixture of sweet potato flour (80%) and groundnut cake flour (GNC) (20%) and (d) a mixture of cassava flour (80%) and GNC (20%) as 25% replacement of the cereal in poor finger millet vegetarian diets. There was no significant difference in weight gain among diets (A), (B) and (C). Diets (D) and (E) containing groundnut cake flour produced significantly higher gains than diet (A).

Subrahmanyan et al (1954b) added alcohol-extracted cottonseed flour to poor finger millet vegetarian diets. Extraction reduced the gossypol content from 2.4% to 0.28%. Diet (A), poor finger millet diet, was compared with diet (B), poor finger millet diet plus 10% cottonseed flour. The improvement in weight gain from the addition of cottonseed flour was highly significant.

Sur et al. (1954a) added food yeast (Torula utilis) to poor vegetarian diets similar to those reported by Murthy et al (1950). In a further series of tests (Sur et al, 1954b), the same samples of finger millet and food yeast were fed in diets at a 5.0% level of protein. The food yeast at the 5.0% protein level improved the feed/grain ratio of the finer millet diets.

Guttikar et al (1968) reported that supplementation of two protein foods to poor vegetarian finger millet diet significantly increased weight gain, haemoglobin and red blood cell count and improved feed/gain ratio. The livers of rats on the basal finger millet diet were significantly lower in protein and higher in fat than those fed the protein enriched diets. The livers of rats fed the poor finger millet diet showed all type of protein deficiency where as the liver sections were normal in rats fed with supplemented diets.

Edible screw-pressed coconut cake from a Mysore mill was compared with SMP in poor Indian finger millet diets (Daniel et al., 1968b). The coconut meal and SMP provided about 2.0% extra protein in the diets. The weight gain on diet (B) finger millet plus minerals and vitamins was not significantly higher than (A) finger millet alone. The inclusion of coconut meal, diets (C) unfortified and (D) fortified significantly improved weight gain over (A) and

(B) but (C) and (D) were not significantly different. The highest weight gain was from fortified SMP, diet (E). The authors suggested that the finger millet was adequate in minerals and vitamins and that coconut meal is an effective protein supplement for children.

Jansen (1974) using FAO (1970) and Orr and Watt (1957) data calculated the amino acid score for lysine, threonine, tryptophan, and methionine plus cystine against whole egg protein for finger millet, supplemented with soy flour (8.1% N), fish meal (12.0% N), groundnut flour (9.4% N) and cottonseed flour (8.0% N) at levels from 0 to 32 parts of supplement/100 parts of finger millet. At all levels of addition of cottonseed flour and groundnut flour lysine was the limiting amino acid. Methionine plus cystine became limiting at 32 parts of soy flour and at 16 and 32 parts of fish meal per 100 parts of finger millet.

Physiological effects of feeding ragi, cowpea, brinjal, guava and fractions of ragi and cowpea starch were studied by Saraswathi et al (1983) in rats. All the materials except ragi starch increased faecal weight with a concomitant decrease in the transit time (TT).

The role of finger millet feeding on skin antioxidant status, nerve growth factor (NGF) production and wound healing parameters in healing impaired early diabetic rats was reported by Rajasekaran et al, (2004). Hyperglycemic rats received food containing 50 g/100 g finger millet (FM). Non-diabetic controls and diabetic controls received balanced nutritive diet. In hyperglycemic rats fed with finger millet diet, the healing process was hastened with an increased rate of wound contraction. In diabetic rats the TBARS levels of both normal and wounded skin tissues were significantly elevated ($P<0.001$) when compared with control (non-diabetic) and diabetics fed with FM. Impaired production of NGF, determined by ELISA, in diabetic rats was improved upon FM feeding. Histological and electron microscopical evaluations revealed the epithelialization, increased synthesis of collagen, activation of fibroblasts and mast cells in FM-fed animals. Thus, increased levels of oxidative stress markers accompanied by decreased levels of antioxidants play a vital role in delaying wound healing in diabetic rats. However, FM feeding to the diabetic animals, for 4 weeks, controlled the glucose levels and improved the antioxidant status, which hastened the dermal wound healing process.

Prashant et al, (2005a) studied the influence of finger millet and kodo millet on rat dermal wound healing was assessed by making a 4 cm2 (2 x 2 cm) excision wound on the shaven back of rats under ether anesthesia. The number of days for complete closure of wounds was lower for finger millet

(13 days) and kodo millet (14 days) treated rats in comparison to untreated (16 days) rats. The results implicate a possible therapeutical role for finger millet and kodo millet in accelerating the process of wound healing.

The beneficial role of a finger millet and kodo millet-based diet in protecting against oxidative stress and maintaining glucose levels in vivo in type II diabetes was investigated Prashant et al (2005b). Whole grain flour of finger millet and KM was incorporated at 55% by weight in the basal diet fed to alloxan-induced diabetic rats over a period of 28 days. Blood glucose, cholesterol, enzymatic and nonenzymatic antioxidants, lipid peroxides in blood plasma, and glycation of tail tendon collagen were measured. The rats fed the KM-enriched diet showed a greater reduction in blood glucose (42%) and cholesterol (27%) than those fed the finger millet (36% and 13%). The levels of enzymatic (glutathione, vitamins E and C) and nonenzymatic antioxidants (superoxide dismutase, catalase, glutathione peroxidase, and glutathione reductase) and lipid peroxides were significantly reduced in diabetic animals and restored to normal levels in the millet-fed groups. Glycation of tail tendon collagen was only 40% in the finger millet–fed rats and 47% in the KM-fed rats compared to the controls. Diets containing whole grain millet meal flour can protect against hyperglycemic and alloxan-induced oxidative stress in Wistar rats.

The effect of aqueous and alcohol extracts of *Eleusine coracana* Linn. on calcium oxalate nephrolithiasis has been studied in male albino rats by Bahuguna et al (2009a). Supplementation with aqueous and alcohol extracts of *E.coracana* grains significantly reduced the elevated urinary oxalate, showing a regulatory action on endogenous oxalates synthesis. The prophylactic and therapeutic treatment with aqueous and alcohol extracts of grains of *E.coracana* had an inhibitory effect of crystal growth, with improvement of kidney function as well as cytoprotective effect.

Bahuguna et al (2009b) reported that the ethanolic extracts act as effective hypernatremic, hyperchloremic and hyperkalemic diuretics (increased Na+, K+ and Ca- excretion volumes). The study justifies that the traditional use of *E.coracana* as diuretic.

The effect of feeding a diet containing 20% finger millet seed coat matter (SCM) was examined in streptozotocin-induced diabetic. The millet SCM feeding showed pronounced ameliorating effects on kidney pathology as reflected by near normal glomerular and tubular structures and lower glomerular filtration rate compared with the shrunken glomerulus, tubular vacuolations in the DC group. Thus, the present animal study evidenced the hypoglycaemic, hypocholesterolemic, nephroprotective and anticataractogenic properties of finger millet SCM, suggesting its utility as a functional ingredient in diets for diabetics. (Shobana et al, 2010)

9.2. Human Studies

Over the ages, a part of the population is using finger miller in their diet, both in Asia and Africa. But to extend the usage to maximum population, though awareness on nutritional aspects is slowly improving, the actual scientific evidence for the health benefits is expected by the consumers. In this direction, the clinical studies with human subjects are being progressed, but has not gained the momentum. The available literature on the finger millet products supplementation in the diet and their role is discussed.

According to the WHO recommendation, the protocol of nutrition studies is the completion of the animal experiments and when those experiments show conclusively that the food is with nutritive strength and free from deleterious factors or safe, then human studies has to be initiated. In the case of finger millet, the animal studies proved to be more adequate in showing positive benefits. Generally, human studies can be taken up with infants for infant foods, growth studies for children with weaning food or supplementary food and balance studies for nursing and expecting mothers, adults of all age groups for assessing the adequacy and efficacy of the diets.

Nitrogen-Balance Studies have been used to determine the adequacy of proteins in meeting the needs of the various age groups and also for assessing the effect of amino acid supplementation in improving the nutritive value of the proteins. Joseph et al (1959) in N, Ca and P metabolism trials with eight girls aged 9-10 years resident in a boarding home in Mysore city employed rice and finger millet of the same protein contents. The children remained in positive nitrogen balance but nitrogen retention was reduced as the proportion of finger millet in the diet was increased. Joseph et al (1958a) reported on the metabolism of N, Ca, and P in eight girls, between 9 and 10 years old, living in a boarding home in Mysore City, fed a poor Indian diet based on finger millet. Finger millet represented about 64% of the total diet, fed during 15 days with collection of urine and faeces over the last 5 days. All the subjects maintained positive N, Ca and P balance. The mean daily N intake was 4.51g and retention 0.52g. Mean daily excretion of N in the faeces was 2.11g, amounting to about 47% of intake; apparent protein digestibility was only about 53%. The mean Ca intake was 1151 mg, retention 226 mg, and mean loss in the faeces 889 mg, about 77% of ingested Ca. Mean Ca retention corresponded to 20% of intake. Mean daily P intake was 887mg, retention 135mg, and mean daily loss in the faeces 634 mg, about 71% of ingested P. Mean P retention corresponded to 15% of intake. As the proportion of finger millet increased, the mean intake and retention of Ca and the mean intake of P increased. Retention of P did not change significantly. Doraiswamy et al. (1969) described the effects of supplementing (A) poor finger millet diets with (B) lysine, (C) leaf protein and (D) low fat sesame flour on the growth over 6 months of 80 boys in a boarding home near Mysore. Nitrogen metabolism was also studied with 6 boys. The lysine added, 0.66g/child/day, diet (B), provided the total 60mg/kg body weight required. Diet (A) was significantly inferior to the other three diets in nitrogen retention and apparent digestibility. Differences in nitrogen retention among diets (B), (C) and (D) were not significant. In apparent digestibility, diet (B) was significantly higher than (A) but lower than (C) and (D) between which the differences were significant.

Growth studies with weaned infants and preschool children for weaning and supplementary foods has been the focus of clinical trials in the recent years. Daniel et al (1965) supplemented (A) a poor finger millet diet with (B) L-lysine, and (C) DL-threonine and together with (D) a skim milk powder diet, compared the digestibility coefficient, biological value of the protein and NPU. The test subjects were 8 girls, 11-12 years of age, living in a boarding home in Mysore city. The diets were fed five 8 day periods, faeces and urine for N determination being collected during the latter 4 days. The order of feeding was: (A) finger millet diet, (B), (C) SMP and (E) low protein diet. The finger millet diet provided 28.7g protein (N × 6.25)/day derived from finger millet 23.5, pigeon peas 3.4g and SMP 1.3g. At the level fed (1.3 g protein/kg body weights) the finger millet was deficient in lysine according to the requirements but was adequate in all other essential amino acids. On diet (A) nitrogen retention was 5.8% of intake. Retention increased significantly in (B) over (A), (C) over (B), and (D) 33% over (C) 24.4%. True digestibility coefficient of (C) was 73% and of (D) 88%. The BV of (D) 85.3 was significantly higher than (B) 81.2 and (B) significantly higher than diet (A). Diet (D) 74.8 was much higher than (B) and (C). Kurein and Doraiswamy (1967) reported that the refined FM flour as compared to whole FM flour fed to 36 boys in the form of dumplings for 5.5 months for lunch and dinner did not show significant difference between the two groups with reference to height, weight, nutritional statistics, reflecting the similar action of refined and unrefined flour in the human system. However, after 4 months, significant increase in protein digestibility of the refined flour diet as compared to the whole grain flour was observed.

Doraiswamy et al. (1969) described the effects of supplementing (A) poor finger millet diets with (B) lysine, (C) leaf protein and (D) low fat sesame flour on the growth over 6 months of 80 boys in a boarding home near Mysore. The lysine added, 0.66g/child/day, diet (B), provided the total 60mg/kg body weight required. The protein and lysine contents of the leaf protein and sesame flour and the effect of the diets on height, weight, haemoglobin and red blood cells were evaluated. By all criteria diets (B), (C) and (D) were significant for height and weight but not for haemoglobin and RBC values. The height and weight from (C) leaf protein were significantly greater than (B) and (C). The general nutritional status was significantly better from diets (B), (C) and (D) than diet (A) but differences among (B), (C) and (D) were not significant. All the diets were lower in calories and diets (A) and (B) lower in protein than recommended by ICMR (1944). The best growth responses were obtained from (C) leaf protein, which provided both additional lysine and protein. Diet (D), sesame flour supplied additional protein but only half the lysine provided by (B) and (C). A weaning food for children aged 6-24 months was formulated by Mbithi et al (2002) using finger millet, kidney beans, peanuts and mango on the basis of amino acid scores and energy (calorie) content. Finger millet and kidney beans were processed by germination, autoclaving and lactic acid fermentation, then blended with roasted peanut paste and mango puree; resulting mixture was dried and milled. A mixture containing 65.2, 19.1, 8.0 and 7.7% DM of processed finger millet, kidney beans, peanuts and mango, respectively, gave a composite protein with in vitro protein digestibility of 90.2% and amino acid chemical score of 0.84. This mixture had an energy density of 16.3 kJ/g of DM, a decreased antinutrient content and a measurable improvement in the in vitro extractability for Ca, Fe and Zn. A 33% (w/v) pap (made from the milled mixture reconstituted with water) had an energy density of 5.4 kJ/g of pap, and it is suggested that this mixture is sufficient to meet the energy requirements of well-nourished children of 6-24 months of age at 3 servings a day and at the FAO average breast-feeding frequency. In an effort to investigate diet in relation to nutrition status of children, diet and dietary intake were investigated in rural Tanzania by Tatala et al (2007). The effect of germination of finger millet based food recipe on its nutritional value was evaluated. The food consisted of finger millet flour, kidney beans, ground peanuts and dried mangoes at predetermined proportions of 75:10:10:5 respectively. Dietary habits of young children were investigated and effects of a fortified food supplement and the cereal based recipe on nutrition status of children were investigated. The two diets were then supplemented to children for 6 months and changes on anaemia and anthropometrical indices of children were evaluated at follow up periods. To assess anaemia and iron status, haemoglobin (Hb), haematocrit (Hct), erythrocyte protoporphyrin (EP) and serum ferritin (SF); and weights and heights were measured to assess growth. A significant improvement in nutrient density was noted in processed cereals. Bioavailability of iron in cereal based diet increased from 0.75±18 to 1.25 ± 41 mg/100 g (P = 008), viscosity was significantly raised by 12% and phytate concentration was reduced from 4.5 ± 0.5 to 4.1 ± 0.5 mg/g (P = 0.03). Significantly lower intake of iron was observed in school children with Hb less than 11.5 g/dl) compared to those who were normal. Total iron intake was 22± 7 and 27 ± 13 mg/day, respectively ($P < 0.05$). There was a significant correlation between iron intake and serum ferritin ($r = 0.233$, $P < 0.05$). After six months of supplementation, children with the fortified beverage had shown significantly larger increase in Hb as compared to the control group. In conclusion, consumption of foods with low iron bioavailability is a major cause of anaemia. Germination improves the nutritional value of foods however there is need to fortify such processed foods for infant feeding.

The impact of ragi based supplementation as mid morning snacks for school children on their nutritional status, biochemical parameters and learning attributes was studied by Devaki et al (2008). Two schools were selected for carrying out the study one which provided mid-day meal, another school which did not provide mid-day meal. A total number of 206 children (10-11 years) studying in fifth and sixth standards from both the schools were selected for the study. Children were fed with mid morning snacks from Aug 2007 to Feb 2008 (152 days excluding holidays). Finger millet based snacks formulated, prepared and supplied by Defence Food Research Laboratory (DFRL), Mysore for feeding children were millet sweetened mix, millet spiced mix and millet sweet cookies. The impact of supplementary feeding in children was measured before the supplementation of snack items and after supplementation in terms of anthropometric measurements (Jelliffe, 1966), dietary intake (Jelliffe, 1966), biochemical attributes in terms of calcium, haemoglobin, ferritin and zinc and learning attributes in terms of intelligence, memory, attention, reasoning and cognition. The Impact of supplementation on nutritional status, biochemical parameters and learning attributes of school children has also been evaluated. In respect of anthropometric measurements, the height of children increased by 1 to 2 cm and weight by 2 Kg. It was found that there was an improvement in both heights and weights after supplementation in all children and was significant at 5% level. The nutritional grades based on the Gomez classification (1955), the children were classified to their nutritional grades, before the supplementation of snack items majority of boys (39%) were mildly malnourished. An improvement was seen in supplementation of snack items in nutritional grades of both boys and girls. The nutritional grades of the Children were shifted to normal by 2 – 4 %. Significant improvement was observed in the nutritional grades of the Children supplemented with snack items. The dietary intake of Children was determined by 24 hour recall method and nutrients such as energy, protein, calcium and iron were computed. Percent adequacy of the nutrient intake with reference to ICMR recommendations was computed. Before supplementation, the dietary intake reveals that energy (84% and 88%), protein (80% and 82%) was adequate, where as calcium (71% and 73%) and iron intake was not adequate (52 - 55%). The nutrient intake of Children was increased after the supplementation of the snack items. After supplementing the mid morning snacks, energy, protein and calcium intake were found to be adequate (80 – 99%). Calcium was not adequate before the supplementation where as it was meeting the adequacy after the supplementation.

Biochemical Attributes with reference to hemoglobin and plasma ferritin, the impact after the supplementation of snack items was considerable. The hemoglobin and ferritin levels improved with a shift in haemoglobin level to next grade and increased ferritin content by 12 – 21 ng/ml of blood. It was observed that majority (76%) of children suffered from the mild form of anemia before supplementation. There was an overall decrease in mild form of anemia (48%) followed by increase in normal level (53%). Perhaps this moderate level of improvement was due to the fact that snacks and their diets provided iron at moderate level of adequacy in terms of RDA (56 to 59%). Moderate form of anemia was also decreased after supplementation. The serum calcium and serum zinc content of Children showed an improvement in calcium level which was due to the ragi based snack which is rich in calcium. The decrease was observed in serum zinc levels after the supplementation. The reason for the decrease in zinc levels may be due to fact that the iron appears to lower zinc absorption. However, boys representing school with school lunch supplemented with ragi based snacks showed an increase in serum zinc levels despite high intake of iron through supplementary snacks.

The learning attributes of children in terms of intelligence level, memory, reasoning- cognition and attention showed an encouraging results. Intelligence as assessed by Raven's Colored Progressive Matrices (Raven, 1956) before and after the supplementation, of millet snacks showed an improvement in intelligence level after supplementation of snack items in the children and it increased from 16 to 23 in 206 numbers of children. Memory as measured by

Pershad Gestalt Inventory test (Pershad and Wig, 1994) showed improved memory level. Reasoning level as per Koh's Block test (Hutt, 1932) showed improved reasoning levels of the Children. Attention level tested by Knox cube method (Stone and Wright, 1980) showed better improvement in their immediate response and their attention in the class. Thus it may be stated that the scores of learning attributes of Children increased after the supplementation of ragi based snack items. Supplementation of mid morning snacks to children who consumed mid day meals at school or food brought from home showed improvement in haemoglobin followed by ferritin level. Children also improved their nutritional status followed by improvement in learning attributes namely intelligence, memory, reasoning and attention. Provision of mid morning snacks helped children to consume more iron and calcium which was reflected in their biochemical indicators. Efforts should be made to introduce mid morning snacks enriched with iron and calcium along with school lunch in order to combat iron deficiency anemia which plays a significant role in school performance. In addition to this, awareness programmes for children and parents should be conducted to enrich their knowledge regarding health and nutrition to visualize better future.

Balance studies with adults and other age group is generally conducted to assess the adequacy of diets with respect to different nutrients. In 1955, Subramanyan et al reported on the metabolism of N, Ca, and P in eight healthy adult males, from 24-32 years old, on a poor Indian diet based on finger millet (variety H 22), over a period of 12 days with the final 5 days used for the collection of urine and faeces. All subjects maintained a positive N, Ca and P balance. Average daily N intake was 10.6g and retention 1.1g. The excretion of N in the faeces was high, 5.5g; apparent protein digestibility was only 50%. Average daily intake of Ca was 3.4g and retention 0.1g. Average daily P intake was 2.3g, retention 0.2g. The average daily excretion of phytate P in the faeces was only 0.3g; about 85% of ingested phytate was hydrolysed.

Fermented products of legumes, Bengal gram dhal (Cicer arietinum) and green gram dhal (*Phaseolus aureus*), and millets, bajra (*Pennisetum typhoideum*), jowar (*Sorghum vulgare*) and ragi (*Eleucine coracana*), were analysed for protein quality and vitamin B content by Sultana and Pasala (1981). The true digestibility of fermented jowar increased significantly ($P < 0.05$) but not that of ragi, bajra and the legume products. Biological value and net protein utilisation of both jowar and ragi products increased significantly on fermentation ($P<0.05$) but not that of bajra. The thiamin and riboflavin contents of both the legume and millet products increased with increase in fermentation time.

The glycaemic response of Indian foods particularly that of South Indian foods were studied by Kavita and Prema (1997). The glycaemic response of three different meals in 20 non-insulin-dependent diabetic volunteers were assessed. Each meal was isocaloric and balanced. Each meal comprised 60% carbohydrate, 20% protein and 20% fat, varying only in the type of carbohydrate. The carbohydrate exchange in each meal (rice, wheat and ragi, which form important staple foods of South Indians) contributed 39% of the total calories of the meal. The wheat-based meal showed the lowest glycaemic response, followed by ragi. Glycaemic effect of 3 millet based foods, khichdi, laddu and baati, was investigated in 10 adult females by Sema and Sarita. (2002). The foods were prepared using finger millet (*Eleusine coracana* var. VL Mandua-149) or barnyard millet (*Echinochloa colona* var. VL Mandira-29) as a base, mixed with legumes and fenugreek seeds. All the foods had a hypoglycaemic effect in the females. Glycaemic index (GI) values of finger millet based foods were 25.53, 34.62 and 36.12 and of barnyard millet based foods, were 27.24, 34.68 and 36.71 for khichdi, laddu and baati, respectively. The GI of the different foods were not significantly different ($P > 0.05$).

According to Asna et al (2006) in the human subjects of ragi dumplings showed higher glycemic response as compared to rice and sorghum, but the glycaemic index did not show significant difference thereby conclusive evidence for the promotion of ragi products as antidiabetic could not be drawn.

A few hypoglycemic foods were formulated by Shobana et al (2007), suitable as dietary supplements to diabetic subjects based on wheat, decorticated finger millet, popped (aralu) and expanded (puri) rice each blended separately with legumes, non-fat dry milk, vegetable oils, spices. A 50-g equivalent carbohydrate portion of the foods in the form of thick porridge was provided to eight healthy adult subjects and the postprandial blood glucose response was determined. The Glycemic Index (GI) values were 55.4±9, 93.4±7, 105±6 and 109±8 for wheat-based, millet-based, aralu-based and puri-based formulations. The study revealed the suitability of wheat-based formulation as a food supplement or as meal replacer in diabetic subjects but the unsuitability of rice-based formulations.

Diabetes is one of the major risk factor in retinopathy and cataract. The prevalence of blindness in India is 15 per 1000 while cataract alone accounts for 80% of this blindness. Finger millet polyphenols (FMP) being a major anti-diabetic and antioxidant component, Chethan et al (2008a) have evaluated them for AR (aldose reductase) inhibiting activity. Phenolic constituents in FMP such as gallic, protocatechuic, p-hydroxy benzoic, p-coumaric, vanillic, syringic, ferulic, trans-cinnamic acids and the quercetin inhibited cataract eye lens effectively, the latter was more potent with an IC (50) of 14.8nM. Structure function analysis revealed that phenolics with OH group at 4th position was important for aldose reductase inhibitory property. Also the presence of neighbouring O-methyl group in phenolics denatured the AR activity. Finger millet seed coat polyphenols has been found to inhibit AR reversibly by non-competitive inhibition. Results thus, provide a stronger evidence for the potentials of FMP in inhibiting cataractogenesis in humans.

Mineral contents and bioavailability of 4 home-made composite meals and 6 commercial ready-to-eat (RTE) foods were assessed. Bioavailability of Fe from composite meals based on 4 staple cereals ranged from 1.5 (finger millet meal) to 4.7% (wheat meal), while that of Zn, from 8.5 (rice meal) to 0.31% (sorghum meal) and of Ca, from 5.1 (wheat meal) to 9.8% (rice meal). Rice and finger millet dishes provided more bioavailable Zn and Ca, while wheat and sorghum dishes contained more bioavailable Fe. Commercial RTE foods provided more bioavailable Zn and Ca than the corresponding whole meals. (Bhavyashree et al., 2009).

Inhibitors of alpha glucosidase and pancreatic amylase play a vital role in the clinical management of postprandial hyperglycemia. Although, powerful synthetic inhibitors are available, natural inhibitors are potentially safer. Phenolics showed strong inhibition towards alpha-glucosidase and pancreatic amylase and the IC50 values were 16.9 and 23.5 mg of phenolics, respectively. The study indicated the therapeutic potentiality of millet phenolics in the management of postprandial hyperglycemia (Shobana et al., 2009).

CONCLUSION

The nutritional strength of finger millet has been evaluated by feeding trials both in rats and humans. The early studies of 1930's reports the nutritive value of white variety is superior to brown type and biological value of finger millet protein as 91.5, superior that of brown variety. The rat feeding trials by the various nutritionists have shown the improvement in weight and protein efficiency ratio when rats were administered finger millet with the combination of skim milk powder, mung beans, chick peas and other legumes, with further supplementation of 2-5% groundnut flour followed by vitamins, minerals and amino acids. The supplementation of cottonseed, food yeast, coconut meal with poor vegetarian finger millet diet increased the protein and rate of weight gain. Rat studies have also shown that haemoglobin improvement from finger millet diet was at par with casein diet. Finger millet diet was beneficial in reducing blood glucose and cholesterol and has a vital role in enhancing the wound healing in diabetic rats.

The human studies in Belgium on the evaluation of nutritional strength of finger millet based complimentary food showed that processed formulation was adequate for 6-24 months children at 3 servings level to overcome the nutritional deficiency. In a Tanzanian study on school children showed that germinated finger millet product improve iron availability to overcome anemia, but fortification is preferable. In school children of India, finger millet diet supplemented with lysine, skim milk powder, leaf protein, sesame flour improved anthropometric measurements and haemoglobin levels. Another study in India with finger millet products namely sweetened millet mix, spiced millet mix and millet sweet cookies supplied by DFRL as mid morning snacks showed improvement in haemoglobin followed ferritin level. Children have improved nutritional status followed by improvement in learning attributes namely intelligence, memory, attention, cognition and reasoning and also improved in serum calcium levels. The present scenario on availability of finger millet products, the awareness of consumers regarding finger millets potential, there is an absolute need for increased clinical studies to bring in the scientific evidence for health benefits. Thus there is lot of scope in this field of research.

BIBLIOGRAPHY

Abstract of papers, (1977). American Chemical Society., 173, AGFD, 62.

Adebowale, K.O., Afolabi, T.A. and. Olu-Owolabi, B.I (2005). Hydrothermal treatment of Finger millet (*Eleusine coracana*) starch. *Food Hydrocolloids.,* 19(6), 974-983.

Agte, V.V., Tarwadi, K.V. and Chiplonkar, S.A. (1999). Phytate degradation during traditional cooking: Significance of the phytic acid profile in cereal based vegetarian meals. *J.Food Comp.Ana.,* 12(3), 161-167.

Anandhi, M. and Rashmikapoor. (2003). Effect of natural and pure culture fermentation of finger millet on zinc availability as predicted from HCl extractability and molar ratios. *JFST,* 40(1), 112-114.

Anderson, E. and Martin, J.H. (1949). World production and consumption of millet and sorghum. *Econ. Bot.,* 3, 265-288.

Anitha Gopal, B. and Muralikrishna, G. (2008). Physico-chemical characteristics of native and pancreatic alpha-amylase digested cereal and finger millet starches. *JFST,* 45(4), 300-304; 27.\

Antony, U., Sripriya, G. and Chandra, T.S. (1996a). Effect of fermentation on the primary nutrients in finger millet (*Eleusine coracana*). *J. Agri. Food. Chem.*, 44(9), 2616-2618.

Antony,U, Sripriya, G. and Chandra, T.S. (1996b). Effect of fermentation on the primary nutrients of finger millet [*Eleusine coracana(1)* Gaertn] varieties grown in Ethiopia. *Ethiop J. Health Sci.,* 19(1), 1-7.

Antony, U. and Chandra, T.S. (1997). Microbial population and biochemical changes in fermentating Finger millet (Eleusine coracana). *World J. of Micro.Biotech.,* 13(5),533-537.

Antony, U. and Chandra, T.S. (1998). Antinutritional reduction and enhancement in protein, starch and amino *acid availability in fermented flavor of finger millet (Eleusine coracana). J. Agri. Food Chem.*, 46 (7), 2578-2582.

Antony, U. and Chandra, T.S. (1999). Enzymatic treatment and use of starters for the nutrients enhancement in fermented flour of red and white varieties of finger millet. *J. Agri. Food Chem.*, 47(5), 2016-2019.

Arthi, A., Urooj, A, and Puttaraj, S. (2003). In-vitro Starch digestibility and nutritional important starch fractions in cereals and their mixtures. *Starch / strake.*, 55, 94-99.

Arya, S.S., Premavalli, K.S. and Parihar, D.B. (1976). Precursors of Carbonyls in chapathies. *J.Food Tech.*, 11, 543-547.

Asna Urooj., Rupashri, K and Shashikala Puttaraj. (2006). Glycaemic responses to finger millet based Indian preparations in non-insulin dependent diabetic and healthy subjects. *JFST*, 43(6), 620-625.

Asp, N. G, Johansson, C, G., Hallmer, H and Siljestrom. (1983). Rapid enzymatic essay of insoluble and soluble dietary fibre. *J.Agr.Food. Chem.*,31, 476-479.

Babu, S. (1976). Effect of germination on folic acid content of Bengalgram and Ragi. *Ind.J.Nui.Diet.*, 13(5), 139-141.

Bahugana, Y.M., Rawat, M.S.M., Juyal, V. and Gnanarajan, G. (2009a). Antilithiatic effect of grains of *Eleusine coracana. Saudi Pharm. J.*, 17(2), 182-188.

Bahugana, Y.M., Mohan Singh. and Maniyri Rawat. (2009b). Diuretic activity of grains of *Eleusine coracana. J.Phar.Res.*, 2(4), 775-776.

Balakrishna Rao, K., Mithyantha, M.S., Devi, I.S. and Perur, N.G. (1973). Nutrient content of some new ragi varieties *J.Agric.Sci., Camb.*, 7, 562-565.

Barbeau, W.E and Hilu, K.W. (1993). Protein, calcium, iron and amino acid content of selected wild and domesticated cultivar finger millet. *Plant Foods for Human Nutrition.*, 43(2), 97-104.

Basappa, S.C. (2002). Investigations on *Chhang* from Finger millet (*Eleucine Coracana* Gaertn)and its commercial prospects. *Ind. Food Industry.*, 21(1), 46-51.

Bender, A.E. and Doell, B.H. (1957). Biological evaluation of proteins: a new aspect. *Brit.J.Nutr.*, 1, 140-145.

Bhavyashree, S.H., Jamuna Prakash., Kalpana Platel., Srinivasan, K (2009). Bioaccessibility of minerals from cereal-based composite meals and ready-to-eat foods. *JFST*, 46(5), 431-435.

Bjork, I., Nyman, M. and Asp, N.G. (1984). Extrusion cooking and dietary fibers Cereal. *Chem.*, 61(2), 174-179.

Bravo, L. (1998). Polyphenols: Chemistry, dietary sources, metabolism and nutritional significance. *Nutr. Rev.,* 56, 317-333.

Brouns, E., Ketlolifnz, B and Arrigoni, E, (2002). Resistant starch and the butyrate revolution. *Trends in Food Science and Technology.,* 13(8), 251-261.

Bullard, R.W. and Holguin, G. (1996). Volatile components of unprocessed rice. *J. Agri. Food Chem.,* 25(1), 99-103.

Camire, M.E. (2001). *Extrusion cooking: Technologies and applications,* edited by Guy R., Woodhead Publishing Co., Cambridge, pp 109-129.

Castelluccio, C., Paganga, G., Melikian, N., Bolwett, G. P., Pridham, J., Sampson, J and Rice-Evans, C. (1995). Antioxidant potential of intermediates in phenylpropanoid metabolism in higher plants. *FEBS Letts.,* 368, 188-192.

Champ, M.M. J (2004). Physiological aspects of resistant starch and in vivo measurements. *Journal of Association of Analytical Chemists International.,* 87 (3), 749-755.

Chandrashekara,M.R. and Swaminathan, M. (1953). Enzymes in ragi and ragi malt. I. amylases. *J.Sci. Ind.Res.,* 13B, 51-54.

Chandrashekara, A and Shahidi, F. (2010). Content of Insoluble bound phenolics in millets and their contribution to antioxidant capacity. *J.Agric. Food Chem.,* 58, 6706-6714.

Chattapadhyaya, S., Ramanathan, M., Das, J. and Bhattacharya, S.K. (1997). Animal models in experimental diabetes mellitus. *Indian J. Exp. Biol.,* 35, 141-145.

Chavan, U. D., Shahidi, F and Naczk, M. (2001). Extraction of condensed tannins from beach pea (*Lathyrus maritmus* L) as affected by different solvents. *Food Chem.,* 75, 509-512.

Chethan, S. and Malleshi, N.G. (2007). Finger millet polyphenols: Characterisation and their nutraceutical potential. *Am. J. Food. Tech.,* 2,582-592.

Chethan, S., Dharmesh, S.M. and Malleshi, N.G. (2008a). Inhibition of aldose reductase from cataracted eye lenses by finger millet *(Eleusine coracana)* polyphenols. *Bioorg.Med.Chem.,* 16, 10085-10090.

Chethan, S., Sreerama, Y.N and Malleshi, N.G (2008b). Mode of inhibition of finger millet malt amylases by the millet phenolics. *Food Chem.* 111(1), 187-191.

Chidda Singh., Prem Singh and Rajbir Singh.(2003). Finger millet p 164-171. In: *Modern Techniques of Raising Field Crops.* 2nd Ed., Oxford & 1BH Publishing Co.Pvt.Ltd., New Delhi.

Chikubu, S (1970). Stale flavor of stored rice. *Jpn. Agric. Res. Quart.,* 63(5), 73-78.

Chitre, R.G., Desai, D.B and Raut, V.V. (1955). The nutritive value of pure bread strains and cereals and pulses. Part I. Thiamine, riboflavin and nicotinic acid contents of 107 pure bread strains of cereals and pulses. *Indian J. Med. Res.,* 43, 575-583.

Chun,J. and HO,C.T.(1999).Comparison of volatile generation in serine/threonine/glutamine-ribose/glucose/fructose model systems. *J. Agric. Food Chem.,* 47, 643-647.

Collins, E. (1971). Steam volatile components of roasted barley. *J.Agri.Food Chem.,* 19, 533-37.

Colonna, P., Gallant, D. and Mercier, C. (1980). Pisum sativum and Vicia faba carbohydrates: Studies of fractions obtained after dry and wet protein extraction process, *Journal of Food Science.,* 45, 1629-1636.

Cranin, A.D. (1982*) Techniques of analysis of flavour in food flavours* part A; Introduction: Eds Morton, I.D., Maclead, A. J, Elsevier Scientific Publishing Co, New York, 15-49.

Cummings, J.H.and Englyst, H.N (1991). Measurement of starch fermentation in the human large intestine. *Canadian Journal of Physiol Pharmacol.,* 69, 121-129.

Dada, L.O., Dendy DAO,B.K. (1988). Cyanide content of germinated cereals and influence of processing techniques. *Int. Dev. Res. Centre.,* 359-365.

Daniel, V.A., Leela, R., Doraiswamy, T.R., Rajalakshmi, D., Venkat Rao, S., Swaminathan, M. and Parpia, H.A.B. (1965). The effect of supplementing a poor Indian ragi diet with L-lysine and DL-threonine on the digestibility coefficient, biological value and net protein utilization of proteins and retention of nitrogen in children. *J.Nutr.Diet.,* 3, 10-14.

Daniel, V.A., Desai, B.L.M., Urs, T.S.R., Rao, S.V., Swaminathan, M. and Parpia, H.A.B. (1968a). The supplementary value of Bengal gram, red gram, soya bean, as compared with skim milk powder to meet poor Indian diets based on ragi, kafficorn and pearl millet. *J.Nutr.Diet.,* 5, 283-291.

Daniel, V.A., Urs, T.S.R., Desai, B.L.M., Rao, S.V., and Swaminathan, M. (1968b). Supplementary value of edible coconut oil meal to poor Indian diets based on rice, ragi, wheat and sorghum. *J.Nutr.Diet.,* 5, 104-109.

Daniel, V.A., Narayanaswamy, D., Desai, B.L.M., Kurein, S., Swaminathan, M. and Parpia, H.A.B. (1970). Supplementary value of varying levels of red gram to poor diets based on rice and ragi. *J.Nutr.Diet.,* 7, 358-362.

Daniel, V.A., Kurein, S., Narayan, D. and Swaminathan, M. (1974). Supplementary relations between proteins of Bengal gram, rice and ragi (*Eleusine coracana*). *Ind.J.Nutr.Diet.*, 11, 137-143.

Decondelle, (1886). *Origin of cultivated plants,* Kegan Paul, Trench & Co., London.

Deepman Sakya. (1989). Cropping systems, Production Technology, pests, diseases of finger millet in Nepal. P275-278. In: *Small millets in Global Agriculture.* Ed: A Seetharam, K.W. Riley and G. Harinarayana. Oxford & IBH Publishing Co. Pvt.Ltd., New Delhi.

Deosthcle, Y. G., Nagarajan, V and Pant, K. C. (1970) Nutrient composition of some varieties of Ragi (*Eleusine coracana*). *Ind. J.Nutr. Diet.,* 7, 80-84.

Desai, A.D., Kulkarni, S.S., Sahoo, A.K., Ranveer, R.C. and Dandage, P.B. (2010). Effect of supplementation of malted ragi flour on the nutritional and sensorial quality characteristics of cake. *Advance JFST,* 2(1), 67-71.

Desai, B.L.M., Narayanaswamy, D., Daniel, V.A., Swaminathan, M. and Parpia, H.A.B. (1970). The improvement of protein value of ragi (*Eleusine coracana)* diet by supplementation with limiting amino acids. *Nutr.Rep.Int.,* 2, 185-191.

Desikachar, H.S.R. (1972). Effect of wet heat treatment on the culinary qualities of ragi. *JFST,* 9(3), 149-150.

Desikachar, H.S.R. (1982). *Technology Options for fermentation weaning foods for the economically weaker segments of population in developing countries.* Food & Nutri. Bull., 4, 57.

Devaki, C.S., Meghana, D. R. & Saraswathi, G. (2008). *Impact of supplementation of snacks (flaxseed and ragi) on anthropometric biochemical and learning attributes of the school children. In*: Project Report, DFRL, Mysore.

Devaraju, B., Mushtari Begum, J., Shemshed Begum and Vidhya, K. (2006). Effect of temperature on physical properties of pasta from finger millet composite flour. *JFST,* 43(4), 341-343.

Donald, A.W., Lindsay, R.C. and Stuibes, D.A. (1978). Isolation and identification of volatile components from wild rice grain. *J.Agri.Food Chem.,* 26 (4), 816-820.

Doraiswamy, T.R., Singh, N. and Daniel, V.A. (1969). Effects of supplementing ragi (*Eleusine coracana)* diets with lysine of leaf protein on the growth and nitrogen metabolism of children. *Br.J.Nutr.,* 23, 737-743.

Dykes, L. and Rooney, L.M. (2006). Sorghum and millet phenols and antioxidants. *J.Cereal.Sci.,* 44, 236-251.

Eliasson. (1996). *Carbohydrates in Foods, Starch- Physicochemical and functional aspects,* Marcel Deckker,inc., Newyork,USA.

Encyclopedia of Food Science and Nutrition. (2003). Vol 6, 3976.

Encyclopedia of Food Science Technology and Nutrition. (1993). Vol 5, 3092

Englyst, H. N, Kingman, S.M and Cummings, J.H (1992). Classification and measurement of nutritionally important starch fractions. *Eur. J. Clin. Nutri.,* 46(2), S33-S50.

Fagerson, I. S.(1969). Thermal Degradation of Carbohydrates – A review. *J.Ag. Fd. Chem,* 17, 747

Faki, E.l., Derikachar, H.S.R., Tareen, J.A.K and Tharanathan, R.N, (1983). Scanning electron microscopy of invivo and in vitro digested starch granules of chick pea, cow pea and horse gram, lebensm.*Wiss. Technol.,* 17, 276.

FAO. (1970). "Amino acid content of foods and biological data on proteins". *FAO Nutrition Studies.* No.24. FAO, Rome.

FAO. (1973). "Composite flour programme". Documentation Package. Vol I, 2nd Ed, reviewed. *Food and Agricultural Industries Service, Agricultural Services Division,* FAO, Rome.

Fasset, D.W. (1973). *Oxalates: In toxicants occurring naturally in foods, National Research Council,* Washington, DC, pp 346-362.

Ferugson, L.R. (2001). *Role of plant polyphenols in genomic stability.* Mutat Res., 475, 89-111.

Filer, L.J. (1977). *Digestion, absorption and metabolism of starch, carbohydrates and health, Hood, Wardrip and Bollenback,* eds, Westport,Conn.Cop., 39.

Gee, J.M., Johnson, I.T., and Lind. (1992). Physiological properties of resistant starch. *Eur. J.Cli.Nutr.,* 46 (suppl.2), S125-S131.

Geeta, R., Virupaksha, T.K and Shadaksharaswamy, M. (1977). Relationship between tannin levels and in vitro protein digestibility in finger millet. J.Agri.Food.Chem., 25, 1101-1104.

Geeta, R., Virupaksha, T.K. and Shadaksharaswamy, M. (1978). Comparison of the protein fractions of finger millet. *Phytochem.,* 17: 1487-1490.

Giri, K.V. (1938). The availability of phosphorous from Indian foodstuffs. *Ind. J. Med. Res.,* 25, 869-877.

Glew, R.S., Chuang Lu-Te, Roberts, J.L. and Glew, R.H. (2008). Amino acid, fatty acid and mineral content of black finger millet (*Eleusine coracana*) cultivated on the Jos plateau of Nigeria. *Food.,* 2(2), 115-118.

Gomez, F., Galvan, R., Cravioto, J. and Frenk, S. (1955). Malnutrition in infancy and childhood, with special reference to Kwashiorkor. *Adv. Pediatr.,* 7, 131-169.

Gomez, M., Moraleja, A., Oliete, B and Ruiz, E. (2010). Effect of fibre size on the quality of fibre enriched layer cakes. *LWT.,* 43, 33-38.

Gopalan, C. (1981). *Carbohydrates in diabetic diet. India: Bulletin of Nutrition Foundation.,* p 3.

Gopalan, C., Ramasastri, B, V and Balasubramanian, S.C. (2000). *Nutritive Value of Indian Foods,* NIN, Hyderabad, India.

Gopalan, C., Ramasastri, B, V and Balasubramanian, S.C. (2004). *Nutritive Value of Indian Foods,* NIN, Hyderabad, India.

Gowri, B.S., Platel, K., Prakash, J and Srinivasan, K., (2001). Influence of amla fruits *(Emblica officinalis)* on the bioavailability of iron from staple cereals and pulses. *Nutr. Res.,* 21(12), 1483-1492.

Gururaj, H. and Krishna, K.R. (1998). *Finger Millet. In: Science of field crop production.* Oxford and IBH Publishing Co., New Delhi.

Guttikar, M.N., Panemangalore, M., Jayaraj, A.P., Rao, M.N. and Swaminathan, M. (1968). Studies on processed protein foods based on blends of groundnut, bengalgram, soyabean and sesame flours and fortified with minerals and vitamins. II. Amino acid composition and nutritive value of the proteins. *J.Nutr.Diet.,* 2, 24-27.

Harding, R.J., Wren, J.J. and Nursten, H.E. (1978). Volatile basic compounds derived from roasted barley. *J. Inst. Brewing.,* 84, 41-43.

Hariharan, K., Desai, B.L.M., Venkat Rao, S., Rajalakshmi.D, Swaminathan, M. and Parpia, H.A.B. (1967). Effect of supplementing poor Indian diets based on wheat, rice and ragi with vitamins, minerals and groundnut flour on the nutritive value of the diets as judged by the growth of albino rats. *J.Nutr.Diet.,* 4, 56-64.

Harinarayana. (1989). Breeding and Varietal Improvement of Small Millets in India. P 59-70. In: *Small Millets in Global Agriculture.* Ed: A Seetharam, K.W. Riley and G. Harinarayana. Oxford & IBH Publishing Co. Pvt.Ltd., New Delhi.

Hegde, B.R. and Linge Gowda, B.K. (1989). Cropping systems and production technology for small millets in India. In: *Small millets in global agriculture,* pp 115-126. Ed: A Seetharam, K.W. Riley and G. Harinarayana. Oxford & IBH Publishing Co. Pvt.Ltd., New Delhi.

Hegde, P. S., Chandrakasan, G and Chandra, T. S. (2002). Inhibition of collagen glycation and cross linking *in vitro* by methanolic extracts of

Finger millet (*Eleusine coracana)* and Kodo millet (*Paspalum scrobiculatum*). *J. Nutr. Biochem.,* 13, 517-521.

Hegde, P.S., Rajasekharan, N.S and Chandra, T.S. (2005). Effects of the antioxidant properties of millet species on oxidative stress and Glycaemic status in alloxan induced rats. *Nut. Res.,* 25, 1109-1120.

Hegsted, D.M. and Chang, Y. (1965). Protein utilization in growing rats. 1. Relative growth index as a bioassay Procedure. *J.Nutr.,* 85, 159.

Hemalatha, S., Platel, K. and Srinivasan, K. (2007). Influence of germination and fermentation on bioaccessibility of zinc and iron from food grains. *Eur.J. Clin.Nut.,* 61(3), 342-348.

Hemamalini, G., Umapathy, K., Padma Rao., Jamuna Rani. and Saraswathi, G. (1980). Nutritional evaluation of sprouted ragi. *Nutrition Reports International.*, 22(2), 271-277.

Hilu, K.W., Dewet, J.M.G. and Seigler, D. (1978). Flavonoid patterns and systematics in Eleusine. *Biochem. Syst. E.coli.,* 6, 247-249.

Houghen, F.W., Quillian, M. A and Curran, W.A. (1971). Head space vapour from cereal grains, *J. Agric. Food Chem.,* 19, 182-187.

Hridlicka, J. and Janicek, G. Formation of carbonyl compounds in toasting of oat flakes. *Sb Vysoke Skoly Chem. Techl.Praze, Postravinarska Tehnl.,* 1964.8(1), 101-106; From CA: 1966, 64, 18303e.

Huang, H., Hartman, T.G. and HO C.T. (1995). Relative reactivities of amino acids in pyrazine formation. *J. Agric. Food Chem.,* 43, 179-184.

Hulse, J.H., Laing, E.M. and Pearson, O.E., (1980). *Sorghum and the Millets: Their composition and nutritive value,* Academic Press, London.

Hutt, M. L (1932). "The Kohs Block-designs test: a revision for clinical practice". *Journal of Applied Psychology.,* 16 (3), 298–307.

Indira, R. and Naik,M.S. (1971). Nutrient composition and protein quality of some minor millets. *Ind.J.Agr.Sci.,* 41, 795-797.

Jani, M., Aunemueller, G and Schildhach, R. (1999). Studies on the effect of millet malt quality on extract recovery. *Brauwelt.,* 139(23), 1062-1065.

Jansen, G, R. (1974). The amino acid fortification of cereals. *In "New Protein Foods".* (A.A.Altschul,Ed),Vol 1A.Technology. 39-120. Academic Press Inc., New York, USA.

Jelliffe Derrick B. (1966) *Assessment of the nutritional status of the community,* WHO, 12.

Joseph, K., Kurien, P.P., Swaminathan,M. and Subramanyan,V. (1958a). Metabolism of nitrogen, calcium and phosphorous in children on a poor vegetarian diet based on ragi (*Eleusine coracana) Fd.Sci.,* 7, 324-325.

Joseph. K., Kurien, P.P. and Swaminathan, M. (1958b). The metabolism of nitrogen, calcium and phosphorus in children on a poor vegetarian diet based on ragi. *Ann.Biochem.exp.Med.* 18, 195-200.

Joseph, K., Kurien, P.P., Swaminathan, M. and Subrahmanyan, V. (1959). The metabolism of nitrogen, calcium and phosphorus in under-nourished children. 5.The effect of partial or complete replacement of rice in poor vegetarian diets by ragi (*Eleusine coracana)* on the metabolism of nitrogen, calcium and phosphorus. *Br.J.Nutr.,* 13, 213-218.

Kadkol, S.B. and Swaminathan, M. (1954).Chemical composition of different varieties of Ragi (*Eleusine coracana) Bull.Cent. Fd.Technol.Res.Inst., Mysore* 4, 12-13.

Kalpana Platel, Eiperspon, S.W. and Srinivasan. K. (2010). Bioaccessible mineral content of malted finger millet (Eleusine coracana), wheat (Triticum aestivum) and barley (Hordum vulgare). *J.Agri.Food.Chem.,* 58(13), 8100-8103.

Kamalanathan, G., Girija, K.A. and Devadas, R.P. (1971). Possibilities of white ragi (*Eleusine coracana)* in human dietary. *Ind. J. Nutr. Diet.,* 8, 315-318.

Kanchana, S. and Kantha, S. Shurpalekar. (1983).*Effect of incorporating ragi (Eleusine Coracana) husk in the diet.,* 33(4), 279-285.

Kannur, S.B., Premavalli, K.S., Arya S.S., Parihar, D.B. and Nath, H. (1974). Studies on flavouring compounds of chapathies – *Carbonyls. JFST.,* 11, 5-12.

Katti, S. V, Sourav Kumar., Malleshi, N.G. (2008). Studies on the effect of milling finger millet in different pulverisers on physicochemical properties of flour. *JFST.,* 45(5), 398-405.

Kavita, M. S. and Prema, L. (1997). Glycaemic Response to Selected Cereal-based South Indian Meals in Non-insulin-dependent Diabetics. *J,Nutri. Envir. Med.,* 7(4), 287-293.

Kavitha, V., Verghese, S., Chitra, G.R., and Prakash, J. (1998): effects of processing, storage time and temperature on the resistant starch of foods. *JFST*, 35(4), 299-304.

Kempanna, C. and Kavallappa, B.N. (1968). Quantitative assessment for nutritive quality of finger millet. *Mysore J. Agric. Sci.,* 2, 324-329.

Khetarpaul, N. and Chauhan, B.M. (1991). Effect of natural fermentation on phytate and polyphenol content and in vitro digestibility of starch and protein of pearl millet. *J.Sci. Food Agric.,* 55, 189-195.

King, L.(2001). Impaired wound healing in patients with diabetes. *Nurs. Stand.*, 15(38), 39-45.

King,E.J. (1951). *"Microanalysis in medical biochemistry".* 2nd ed. J.A.Churchill Ltd., London.

Krishnamurthy, K. (1968). Nutritive content of ragi varieties in relation to fertilizer levels. *J.Nut.Diet.,* 5, 10-12.

Kumaraswamy, K and Venkataramana, C. R. (1974). Pot Culture studies on the response of CO-10 finger millet to phosphorus fertilization. *Ind.J.Agri.Sci.,* 44, 50-54.

Kuppuswamy, S., Joseph, K., Rao, M.N., Rao, G.R., Sankaran, A.N. Swaminathan, M. and Subrahmanyan, V. (1957). Supplementary value of Indian multipurpose flour to poor vegetarian diets based on different cereals and millets. *Fd.Sci.,* 6, 84-86.

Kurien, P. P and Doraiswamy, T.R. (1967). Nutritive value of refined ragi *(Eleusine coracana)* floir. II. Effect of replacing cereal in a poor diet with whole or refined ragi flour on the nutritional status and metabolism of nitrogen, calcium and phosphorus in children (boys). *J.Nutr.Diet.,* 4, 102-109.

Kurien, S., Daniel, V.A., Venkat Rao, S., Swaminathan, M. and Parpia, H.A.B. (1969). Effect of calorie restriction on the supplementary value of protein foods to poor vegetarian diets based on rice and ragi. *J.Nutr.Diet.,* 6, 111-114.

Lam, H.S. and Proctor, A. (2003). Lipid hydrolysis and oxidation on the surface of milled rice. *JAOCS.,* 80(6), 563-567.

Lasekan, O.O. and Lasekan, W.O. (2000). Volatile Flavour compounds in Kunun Zaki-A Nigerian millet based beverage. *J. Food Quality.,* 23, 185-193.

Lawal, O S (2009). Starch hydroxyalkylation: physicochemical properties and enzymatic digestibility of native and hydroxypropylated finger millet (Eleusine coracana) starch. *Food-Hydrocolloids.,* 23(2), 415-425.

Lee, C.Y., (1977) Pyrazine compounds in canned sweet corn flavor. *Food Chem.,* 25, 459-63.

Leela,R.,Daniel,V.A.,Rao,S.V.,Hariharan,K., Rajalakshmi,D., Swaminathan,M and Parpia,H.A.B. (1965). Amino acid supplementation of proteins.I.The effect of supplementing ragi *(Eleusine coracana)* and ragi diets with lysine, threonine, and skim milk powder on the nutritive value of their proteins. *J.Nutr, Diet.,* 2, 78-82.

Libert, B. and Franceschi, V.R. (1987). Oxalate in crop plants. *J. Agric. Food Chem.,* 35, 926.

Lopez, H.W., Levrat, M.A., Guy, C., Messanger, A., Demigne, C. and Remesy, C. (1999). Effects of soluble corn bran arabinozylans on digestion, lipid metabolism and mineral balance in rats. *J. Nutr. Biochem.,* 10, 500-509.

Lorri, W and Svanberg, U. (1993a). Lactic fermented cereal gruels with improved in vitro protein digestibility. *Int. J. Fd.Sci & Nutr.,* 44(1), 29-36.

Lorri,W. and Svanberg,U. (1993b). Lactic-fermented gruels: Viscosity and flour concentration. *Int.J. of Food. Sci., & Nut.,* 44 (3), 207-213.

Macfarlene, M., Cummings, J.H and Bane, G.T. (1991). The control and conseguences of bacterial fermentation in human colon. *Journal Applied Bacteriology.,* 70, 443-459.

Maga, J.A. (1976). Lactones in Foods. *Cri.Rev.Food Sci. & Nui.*, 8, 1-56.

Maga, J.A. (1978). Cereal volatiles. A Review, *J. of Agri. & Fd. Chem.,* 26(1), 175-178.

Maga, J.A. (1982). Pyrazines in Foods – An update. *Rev. Food Sc. & nui.,* 16, 1-115.

Mahadevappa, V.G. and Raina, P.L. (1978). Lipid profile and fatty acid composition of finger millet. *JFST,* 15,100-102.

Mahadevappa, V.G. and Raina, P.L. (1984). Glyceroglycolipids and glycerophosphatides in finger millet seeds (*Eleusine coracana). JFST,* 21, 268-271.

Mahudeswaran, K., Natarajan, M. and Ramachandran, A. (1966). White ragi – a source for more protein. *Madras. Agric.J.,* 53, 179-180.

Mahudeswaram, K and Ayyam Pernmal, A. (1970). A note on the vitamin content of ragi, *Madras Agric.J.,* 57, 289-290.

Majumdar, T.K., Premavalli, K.S and Bawa.A.S. (2006). Effect of puffing on calcium and iron contents of ragi varieties and their utilization. *JFST,* 43(5), 542-543.

Makokha, A.O., Oniang, O.R.K., Njoroge, S.M and Kanar, O.K. (2002). Effect of traditional fermentation and malting on phytic acid and mineral availability from Sorghum (*Sorghum bicolor*) and finger millet *(Eleusine coracana)* grain varieties grown in Kenya. *Food and Nutrition Bulletin.,* 23(3), Suppl. 241-245.

Malleshi, N.G. and Desikachar,H.S.R. (1979). Malting qualities of new varieties of Ragi. *JFST.,* 16, 149-151.

Malleshi, N.G. and Desikachar.H.S.R.(1981). Varietal differences in puffing quality of ragi (*Eleusine coracana). JFST,* 18, 30-32.

Malleshi, N.G. and Desikachar,H.S.R.(1982). Formulation of a weaning food with low hot paste viscosity based on malted ragi (*Eleusine coracana*) and green gram (*Phaseolus radiatus*). *JFST,* 19, 193-197.

Malleshi, N. G., Desikachar, H. S. R. and Venkat Rao, S. (1986). Protein quality evaluation of a weaning food based on malted ragi and greengram, *Plants Food Hum. Nutri.,* 36(3), 223-230.

Malleshi, N.G. and Amla, B.L. (1988). Malted weaning foods in India. 340-348p. *Proceedings of a workshop held in Nairobi Kenya.,* 12-16, Oct., 1987.

Malleshi, N. G., Balasubramanyam,N.,Indiramma, A.R., Baldev Raj and Desikachar, H. S. R.(1989). Packaging and storage studies on malted ragi and greengram based weaning food. *JFST,* 26 (2),68-71.

Malleshi,N.G., Nirmala, A. Hadimani., Rangaswami Chinnaswamy and Carol, F. Klopfenstein. (1996). Physical and nutritional qualities of extruded weaning foods containing sorghum, pearl millet, or finger millet blended with mung beans and nonfat dried milk. *Plant Foods for Human Nutrition.,* 49(3), 181-189.

Malleshi, N.G and Klopfenstein, C.F. (1998). Nutrient composition and amino acid contents of malted sorghum, pearl millet and finger millet and their milling fractions. *JFST*, (1999), 35(3), 247-249.

Malleshi, N.G. (2003).Decorticated finger millet (*Eleusine coracana*). and process for its preparation. *Patent No. WO 03/080248.*

Malleshi, N.G., Reddy, P.V and Klopfenstein, C.F. (2004). Milling trials of sorghum, pearl millet and finger millet in quadrumat junior mill and experimental roll stands and the nutrient composition of milling fractions. *JFST,* 41(6), 618-611.

Malleshi,N.G. and Singh, U.R. (2005). Process for preparation of finger millet. *Patent No:* WO 2005/063048.

Mamiro, P.R.S., Camp, J. (*Eleusine coracana*). Mvikya, S.M. and Huyghebaert, A. (2001). In vitro extractability of calcium, iron, zinc in finger millet and kidney beans during processing. *J.Food.Sci.,* 66(9), 1271-1275.

Mamiro, P.R.S., Kolsteren, P.W., Camp, J.H.V., Roberfriod, A., Tatala, S., Opsomer A.S. (2004). Processed complimentary food does not improve growth and hemoglobin status of rural Tanzanian infants from 6-12 months of age in Kilosa district, Tanzania. *J. Nutri.,* 134,(5), 1084-1090.

Mangala, S.L., Ramesh, H.P., Udayasankar, K. and Tharanathan, R.N. (1997): Resistant starch derived from processed ragi (Finger millet, *Eleusine coracona*) flour: Structural characterization, *Food Chem.,* 64, 475-479.

Mangala, S. L. Udayasankar, K and. Tharanathan, R. N (1999a). Resistant starch from processed cereals: the influence of amylopectin and non-carbohydrate constituents in its formation. *Food Chem.,* 16, 391-396.

Mangala,S.L., Malleshi,N.G. and Mahadevamma Tharanathan, R N. (1999b) Resistant starch from differently processed rice and ragi (finger millet). *Eur. Food Res. Tech,* 209(1), 32-37.

Mason, E.D., Devadas, R. and Frimodt-Moller, J. (1946). The effects on growth in rats of butter and ragi (*Eleusine coracana)*, separately and combined, as supplements to the poor rice diet of South India. *Ind.J.med.Res.,* 34, 45-58.

Matuschek, E., Towo, E and Svanbergu. (2001). Oxidation of polyphenols in phytate reduced high tannin cereals: Effect on different phenolic groups and on in vitro accessible iron. *J.Agric. Food Chem.,* 49, 5630-5638.

Mbithi Mwikya, S., Ooghe, W., Camp, J.V., Ngundi, D., Huyghebaert, A. (2000a). Amino acids profiles after sprouting, autoclaving and lactic acid fermentation of finger millet (Eleusine coracana) and Kidney beans (Phascolus vulgaris). *J. Agri. Food. Chem.,* 48(8), 3081-3085.

Mbithi-Mwikya, S. Camp, J.V, Yiru, Huyghebacrt, A. (2000b). Nutrient and antinutrient changes in finger millet (*Eleusine coracana*) during sprouting. Lebensmittel-Wissenschaft Und. *Technologie.,* 33(1), 9-14.

Mbithi, Mwikya, S., Camp, J.Van., Mamiro, P.R.S., Ooghe, W., Kolsteren, P. and Huyghebaert,A. (2002). Evaluation of the nutritional characteristics of a finger millet based complementary food. *J.Agrl.Fd.Chem.,* 50(10), 3030-3036.

McKeown, N.M. (2002). Whole grain intake associated with metabolic risk factors for type 2 diabetes and cardiovascular disease. *Am.J.Clin.Nutr.,* 76, 390-398.

McWilliams, M and Mackey, A.C. (1969). Wheat flavor components. *J.Fd. Sci.,* 34, 493-498.

Mitchell, H.H. (1923/24). A method of determining the biological value of protein. *J.Biol.Chem.* 58, 873-903.

Mohan, B.H., Anitha, G., Malleshi, N.G., and Tharanathan, R.N (2005): Characteristics of native and enzymatically hydrolysed ragi and rice starches. *Carbohydrate Polymers.,* 59, 43-50.

Monnier, V.M. (1990). Nonenzymatic glycosylation, the Maillard reaction and the aging process. *J.Gerontol.,* 45, 105-111.

Moruzzi. (1991). Chemistry of some food products from Tripolitania. *Biochem. Terap. Sperim.,* 18(7), 316-319.

Mtebe,K., Ndabikunze, B.K., Bangu, N.T.A. and Mwemezi,E. (1993). Effect of cereal germination on the energy density of togwa. *Int.J.Food Sci.Nuit.,* 44(3), 175-180.

Mugula, J.K and Lyimo, M. (1999). Evaluation of the nutritional quality and acceptability. Weaning foods in Tanzania. *Int.J.Fd. Sci. Tech.,* 50(4), 275-282.

Mugula, J.K., Nnko, S.A.M., Narvhus, J.A and Sørhaug, T. (2002). Microbiological and fermentation characteristics of *togwa*, a Tanzanian fermented food. *Int.J.Food Micr.,* 80, 187-199.

Muir, J.G. (1999). Location of colonic fermentation events, importance of consuming resistant starch with dietary fiber, *Asia Pacific Journal of Clinical Nutrition.,* 8, (S) 14-21.

Muroki, N, M., Maritim, G, K., Karuri, E, G., Tolong, H, K, Imungi, J, K., Kogi Makau, W., Maman, S., Carter, E and MArtezki, A, N. (1997). Nutritionally improved weaning foods. *J.Nutr.Edu.,* 29(6), 335-342.

Murthy, H. B.N., Swaminathan, M and Subrahmanyan, V. (1950). Supplementary value of groundnut cake to tapioca and sweet potato. *J.Scient.Ind.Res.,*9B, 173-176.

Mwanbene, R.O.F. (1989). Finger millet research in the Southern highlands of Tanzania. *Oxford and IBH Publishing Co.Pvt.LTd.* Chapter 16, 155-159.

Mwanbene, R.O.F.(1989). Finger millet cropping systems and management practices in Tanzania. pp 315-320. In*: Small millets in Global Agriculture.* Ed: A Seetharam, K.W. Riley and G. Harinarayana. Oxford & IBH Publishing Co. Pvt.Ltd., New Delhi.

Mwesigye, R. K and Okurut, T.O. (1955). A survey of the production and consumption of traditional alcoholic beverages in Uganda. *Process Biochem.,* 30(6), 497-501.

Narasinga Rao, B.S. and Prabhavathi, T. (1978). An in vitro method for predicting the bioavailability of iron from foods. *Am.J.Clin.Nut.,* 31, 169-172.

Narayan Rao, M., Chandrashekar, M.R., Swaminathan, M and Subramanyan, V. (1958). Manufacture of malt and malt products. *Food Sci.* (Mysore), 7, 57-59.

Narayanaswamy, D., Kurien, S., Desai, B.L.M., Daniel, V.A., Venkat Rao, S., Swaminathan, M. nad Parpia, H.A.B. (1970). Supplementary value of yeast grown on petroleum hydrocarbons to poor diets based on rice and ragi (*Eleusine coracana*). *Nutr.Rep.Int.,* 1, 305-312.

Narayanaswamy, D., Kurein, S., Daniel, V.A., Swaminathan, M. and Parpia, H.A.B. (1972). Effect of incorporation of a low cost protein food (Bal-ahar) in poor rice and ragi diets on their overall nutritive value. *Indian.J.Nutr.Diet.,* 9, 73-77.

Narayanaswamy. D., Daniel, V.A., Kurein, S., Rajalakshmi, D. and Swaminathan.M. (1974). Supplementary value of protein enriched cereal foods containing varying amounts of proteins to poor rice and ragi diets. *Ind.J.Nutr.Diet.,* 11, 72-76.

National Research Council. (1996). Finger millet (Chapter-2). In lost crops of Africa Vol. 1 *Grains Board of Science and Technology for International Development.* National Academy Press, Washington DC.

Navitha, G. and Sumathi, P. (1992): Effect of primary processing on dietary fiber profile of selected millets. *JFST,* 29(5), 314-315.

Nirmala, M., Subba Rao, M.V.S.S.T, Muralikrishna, G. (2000). Carbohydrates and their degrading enzymes from native and malted finger millet (*Eleusine coracana*) (ragi, Indaf15). *Food.Chem.,* 69(2), 175-180.

Nirmala, M. and Muralikrishna, G. (2002). Changes in starch during malting of finger millet (Ragi, Eleusine coracana, Indaf15) and its in vitro digestibility studies using purified ragi amylases. *Eur. Food Res. Tech..,* 215 (4), 327-333.

Niyogi, S.P., Narayana N. and Desai, B.G. (1934) The nutritive values of Indian vegetable foodstuffs V. The nutritive value of ragi. *Ind.J.Med.Res.,* 22, 373-382.

Noda, T., Uda, T. S, S., Mori, Takigawa, S., Matsuura Endo, C., Saito, K., Mangalika, W.H.A., Hanaoka, A., Suzuki, Y and Yamauchi, H. (2004). The effect of harvest dates on the starch properties of various to cultivars, *Food Chem.,* 86, 119-125.

Nont, M.J.R. and Davis, B.J. (1982). Malting characteristics of finger millet, Sorghum and Barley. *J. Inst. Brew.,* 88, 157-160.

Oberleas, D. (1973). Phytates. In *"Toxicants occurring naturally in foods".* Food. Nutri., Bd.Nat.Res.Coun.pp., 363-371. National Academy of Sciences.

Oi, Y. and Kitabatake, N. (2003) Analysis of the carbohydrates in an East African traditional beverage, togwa. *J.Agric. Food Chem.,*51(24), 7029-7033.

Ojijo, N. K.O. and Shimoni, E (2004).Rheological properties of fermented finger millet (Eleucinecoracana) thin porridge. *Carbohydrate-Polymers.,* 57(2), 135-143, IS:

Onyango, C., Okoth, M.W., Mbugua, S.K. (2000). Effect of drying lactic fermented uji (an East African Sour porridge) on some carboxylic acids. *J. Sci. Food & Agri.,* 80(13), 1854-1858.

Onyango, C, Bley, T, Raddatz, H and Henle, T. (2004a). Flavour compounds in backslop fermented uji. *Eur. Food Res Tech.,* 218(6), 579-583.

Onyango, C., Noetzold, H., Bley, T and Henle, T. (2004b). Proximate composition and digestibility of fermented and extruded *uji* from maize-finger millet blend. *LWT.*, 38, 827-832.

Onyango, C., Noetzold, H., Ziems, A., Hofmann,T, Bley T., Henle,T. (2005). Digestibility and antinutrient properties of acidified and extruded maize finger millet blend in the production of uji. *LWT.*, 38(7), 697-707.

Orr, M.L. and Watt, B, K. (1957). Amino acid content of foods. *Home Economics Research Report No.4.* United States Department of Agriculture, Washington, DC.

Passmore, R. and Sundararajan, A.R. (1941). The vitamin B1 content of the millets *Eleusine coracana* and *Sorghum vulgare,* whole wheat grown under manurial conditions and rice stored underground. *Ind.J.Med.Res.,* 29, 89-94.

Patrick, K, Mwesigye & Tom Okia Okurut. (1995). A survey of the production and consumption of traditional alcoholic beverages in Uganda. *Proc,Biochem..,* 30(6), 497-501.

Patwardhan, V.N. (1961). Nutritive value of cereal and pulse proteins. *In "Progress in meeting protein needs of infants and preschool children"*- Publication No.843, pp.201-210.

Periago, M. J., Englyst, H. N and Hudson, G. J. (1996). The influence of thermal processing on the non-starch polysaccharide (NSP) content and *in vitro* digestibility of starch in peas (*Pisum sativum* L). *LWT.,* 29 (1), 33-40.

Pershad, D and Wig, N. N. (1994). *Revised Manual for PGI memory scale* (PGIMS). Agra, Uttar Pradesh, India: National Psychological Corporation.

Perumal, K.R. (1982). Genotypic variation in photosynthetic efficiency and translocation and its relation to leaf character, growth and productivity in finger millet. *Ph.D. Thesis.*

Peter, J., Michelin, I., Tlaskalova, H., Capouchova, I., Famera, O., Urminska, D., Tuckova, L and Knoblochova, H. (2003) *Czech J.Sci..,* 21(2) 59-70.

Platel, S. and Shurpalekar, K.S (1994): Resistant starch content of Indian foods, *Plant Foods for Human Nutrition.,* 45 (1), 91-95.

Pore, M.S. and Magar, N.G. (1977). Nutritive value of hybrid varieties of finger millet. *Ind. J. of Agric. Sciences.,* 47(5): 226-228.

Pore, M.S. and Magar, N.G. (1979). Nutritive composition of hybrid varieties of finger millet. *Ind. J. of Agric. Sciences.,* 49(7): 526-532.

Prachi Gupta and Premavalli, K.S. (2011). Invitro studies on Functional properties of selected natural dietary fibers. *Int.J.Fd.Prop.,* 14, 397-410.

Prashant, S. Hegde., Anitha, B., Chandra, T.S. (2005a) In vivo effect of whole grain flour of finger millet (*Eleusine coracana*) and kodo millet (*Paspalum scrobiculatum*) on rat dermal wound healing. *Indian Journal of Experimental Biology*. 43(3), 254-258.

Prashant, S.H., Rajasekaran.N.S., Chandra, T.S. (2005b). *Effects of the antioxidant properties of millet species on oxidative stress and Glycaemic status in alloxan-induced rats.* 25(12), 1109-1120

Premavalli, K. S., Leela, R, K., Arya, S. S., Parihar, D, B and Nath, H. (1973). Studies on packaging andstorage of Wheat Flour (Atta) under Tropical conditions. II. *JFST.,* 10(1), 27-30.

Premavalli, K.S. and Arya, S.S (1983). Physical and chemical changes during roasting of semolina. *J. Food Tech.,* 11, 469-479.

Premavalli, K. S., Majumdar, T. K., Madhura, C. V. and Bawa, A.S. (2003). Development of Traditional products V.Ragi Based Convenience mixes. *JFST.,* 40(4), 361-365.

Premavalli, K.S. and Bawa, A.S. (2004a). Chemistry of finger millet. *Trends in Carbohydrate Chem.,* 9, 105-108.

Premavalli, K.S. Roopa, S. and Bawa, A.S. (2004b). Effect of variety and processing on the carbohydrate profile of finger millet. *Trends in Carbohydrate Chem.,* 9, 109-113.

Premavalli, K.S., Jagananth, J.H., Majumdhar, T.K. and Bawa, A.S. (2005). Studies on phase transition in ragi (*Eleusine coracona*) starch in relation to gelatinization. *Journal of Food Science and Technology.,* 42 (2), 336-340.

Premavalli, K.S., Satyanarayana Swamy, Y.S., Madhura, C.V., Majumdar, T.K. and Bawa,A.S. (2005). Effect of pre-treatments on the physic-chemical properties of puffed ragi (Eleusine coracana) flour. *JFST,* 42(5), 443-445.

Premavalli, K.S. Roopa, S., Wadikar, D.D., Satyanarayana Swamy, Y.S., Madhura, C.V., Majumdar, T.K., Narayan Prasad, N. and Bawa, A.S. (2006). Major Nutrients of Ragi varieties. *Project Technical Report* Part III DFRL No.145B.

Premavalli, K. S and Shobha, B. (2006). Flavour volatiles of ragi varieties. *ProjectTechnical Report,* Part V,DFRL,145D

Premavalli,K.S. and Roopa, S. (2006). Resistant starch-a functional fibre. Ind. Food Ind., 25(2), 40-45.

Premavalli,K.S. Majumdar,T. K. Wadikar,D.D.,Madhura,C.V., Swamy, Y.S.S., Vasudish C.R., Mohankumar,B.L., Roopa,S,Shoba,B. and Bawa,

A.S. (2006) *Standardization of ragi based Traditional Food Products.* Project Technical Report Part I., 145A,DFRL,Mysore.

Rachie, K.O. and Peters, L.V. (1977). The eleusines: A review of the world literature, *Int. Crops Research Institute for the Semi Arid Tropics,* Hyderabad, India.

Radhika, C., Sarita, S. (2008). Genotype variations in physical, nutritional and sensory quality of popped grains amber and dark genotypes of finger millet. *JFST,* 45(5), 443-446.

Rajalakshmi, P and Geervani. (1990). Studies on tribal Foods of South India: Effect of Processing methods on the Vitamin and *in vitro* Protein Digestibility (IVPD) of Cereals/Millets and Legumes. *JFST,* 27 (5), 260-263.

Rajammal,P.D., Usha C. and Prema,K.S. (1980). Nutritional evaluation of the supplementary value of low-cost and locally available foods to a poor rice or ragi diet: IV. Calcium utilisation of five ragi-based combinations. *Plant Foods Hum. Nutr.*, 30(1), 53-60.

Rajasekaran,N. S., Nithya,M., Rose,C., Chandra, T. S. (2004). The Effect of Finger Millet Feeding on the Early Responses during the Process of Wound Healing in Diabetic rats. *Biochimica et biophysica acta.* 1689(3),190-201.

Ramiah, P.V and Satyanarayana, P. (1936). The quality of crops. I. Nutritive values of different varieties of ragi grains. (*Eleusine coracana). Proc.Ass.Econ.Biol.*, 4, 13-31. (Chem.Abstr.32, 6299).

Ramulu, P and Udayasekhara Rao,P, (1997). Effect of processing on dietary fiber content of cereals and pulses. *Plant Foods. Hum. Nutr.,* 50(3), 249-257.

Rao and Mushenga. (1985). Traditional Food Crops in Zimbabwe. Finger millet. *Zimbabwe Agri.J.,* 82(3), 101-104.

Raven, J. C. (1956). Progressive Matrices, sets A, B, C, D and E. London: H. K. Lewis.

Ravindran G. (1991). Studies on Millets: Proximate composition, mineral composition and phytate and oxalate contents. *Food Chem.,* 39, 99-107.

Ravindran, G. (1992).Seed protein of millet amino acids composition, proteinase inhibitors and in-vitro protein digestibility. *Food Chem.,* 44(1), 13-17.

Ravishankar, C. R and Prakash, C (2008). *Improved agricultural practices for high yielding crops.* UAS, Bangalore, ZARC Publishers

Reddy, N.R, Sathe, S.K., and Salunkhe, O.K., (1982). Phytates in legumes and cereals. *Adv. Food Res.,* 28, 1-21.

Reddy, N.R. and Pierson, M.D. (1994). Reduction in Antinutritional and toxic components in plant foods by fermentation. *Food Res. Int.,* 27, 281-290.

Reinhold, J.G, Ismail Beigi, F. and Faradji,B (1975). Fibre versus phytate as determinant of the availability of Ca,Zn. and Fe. Of breadstuffs. *Nutr. Rpt.Int.,* 12, 75-78.

Rhon, S., Rawel, H.M. and Kroll, J. (2002). Inhibitory effects of plant phenols on the activity of selected enzymes. *J.Agri.Food Chem.,* 50, 3566-3571.

Rooney, W.L., Salem, A. and Johnson, A.J. (1967). Studies on the carbonyl compounds produced by sugar-amino acid reactions in Model systems. *Cereal Chem.,* 44, 576-83.

Roopa, M.R., Urooj, A and Puttaraj, S, (1998): Rate of In-vitro starch hydrolysis and digestibility index of ragi-based preparations. *JFST,* 35(2), 138-142.

Roopa, S. (2006). Studies on starch fractions of Finger Millet and its functionality in processed foods. Ph.D Thesis, University of Mysore.

Roopa, S. and Premavalli, K.S. (2008), Effect of processing on starch fractions in different varieties of finger millet. *Food Chem.,* 106(3), 875-882

Roos, Y. R. J (1995). *Phase transitions in foods,* Academic press, San diego, 15-18.

Sagum, R. Arcot, J. (2000). Effect of domestic processing methods on the starch: Non-starch polysaccharides and in-vitro starch and protein digestibility of 3 varieties of rice with varying levels of amylase, *Food Chemistry.,* 70, 107-111.

Sai Kumar, R S., Annapurna Singh, S. and Appu Rao A G. (2005) Thermal stability of alpha-amylase from malted jowar (Sorghum bicolor). *Journal-of-Agricultural-and-Food-Chem.,* 53(17), 6883-6888;

Saldivar, S. (2003), Cereals: Dietary importance. In: Caballero B, Trugo L, Finglas P (ed) *Encyclopedia of Food Sciences and Nutrition,* Reino Unido: Academic Press, Agosto, London, pp 1027 -1033.

Samantaray, G.T., Misra, P.K., Patnaik,K.K. (1989). Mineral composition of ragi. *J. Nut. Diet.,* 26(4), 113-116.

Samantary, G.T and Samantaray, B.K (1997): X-ray diffraction of ragi (Eleusine coracona). *JFST.,* 34 (4), 343-344.

Sangita and Sarita. (2000). Nutritive value of malted flours of finger millet genotypes and their use in the preparation of burfi. *JFST.*, 37(4): 419-422.

Sankara Rao,D.S. and Deosthale, Y.G. (1983). Mineral composition, ionizable iron and soluble zinc in malted grains of pearl millet and ragi. *Food Chem.,* 11, 217-223.

Sankara V, A., Joseph, R. and Meenakshi Ganesan, N. (1998). Genetic variability and diversity for protein and calcium contents in finger millet (*Eleusine coracana)* L Gaertn in relation to grain colour. *Plant Food Hum. Nutr.,* 52(4): 353-364.

Saraswathi, G., Sundaravalli, O. E. and Kantha, S. S. (1983). Physiological effects of dietary fibre of some Indian foods in rats. *Plant Foods for Human Nutrition*, 33(4), 243-249

Sarita, S. and Anju, B (1998). Popping qualities of minor millets and their relationship with grain physical properties. *JFST,* 35(3), 265-267.

Scheppack, E, Fabian, C, Sachs, M and Kaspar H (1988). Effects of starch malabsorption on fecal SCFA excretion in man sand. *Journal of Gastro-enteral.*, 23,755-759.

Seetharam, A. (2001). *Technology for increasing production of Finger millet and small millets in India.* All India Coordinated Small Improvements Project ICMR, UAS, GKVK, Bangalore.

Selvaraj, A., Balasubrahmanyam, N. and Haridas Rao, P. (2002). Packaging and storage on biscuits containing finger millet (ragi) flour. *JFST.,* 39(1), 66-68.

Sema, A. and Sarita, S. (2002). Suitability of millet-based food products for diabetics. *JFST,* 39(4), 423-426.

Shahidi, F., Janitha, P. K and Wanasundara, P. D. (1992). Phenolic antioxidants. *Crit. Rev. Food. Sci. Nutr.,* 32, 67-103.

Shankara, P. (1991). *Investigations on preharvest and post harvest aspects of finger millet.* Ph.D. Thesis, University of Mysore, Mysore.

Sharavathy, M.K., Urooj, A., Puttaraj, S. (2001). Nutritionally important starch fractions in cereal based Indian Food preparations. *Food Chem.,* 75, 241-247.

Shashi, B.K., Sunanda, S., Shailaja, H., Shankar, A.G. and Nagarathna, T.K. (2007). Micronutrient composition, Antinutritional factors and bioaccessibility of iron in different finger millet (*Eleusine coracana*) genotypes Karnataka *J.Agric.Sci.,* 20(3), 583-585.

Shashidar, V.R., Gurumurthy, B.R., Prasad, T.G., Mudayakumar, A. Seetharam and. Krishnasastry, K.S. (1984). Genotypic variation in carbon exchange rate, functional leaf area and productivity in finger millet (*Eleusine coracana*). *Field Crops Research.,* 13, 33-146.

Shayo, N.B., Tiisekura, B.P.M., Laswai, H.S. and Kimaro, J.R. (2001). Malting characteristics of Tanzania finger millet varieties. *Food Nutr. J. Tanzania.,* 10(1), 1-4.

Shepherd, A.D., Woodhead, A.H. and Kapasi, Kakama J. (1971-72). Cereal processing in 'Annual Report 71-72, *East African Industrial Research Organisation.,* pp 23-52.

Shibamato, T. and Bernhard, R.A. (1977). Ryrazine formation in model systems. *Agric. Biol.Chem.,* 41, 143-44.

Shigematsu, H., Kurata, T., Kato, H. and Fujimaki M (1972). Volatile compounds formed on roasting DL α-alanine with D-glucose. *Agric. Biol. Chem.,* 36, 1631-35.

Shimelis Admassu., Mulugeta Teamir., Dawit Alemu. (2009). Chemical composition of local and improved finger millet (*Eleusine coracana L. Gaertin*) varieties grown in Ethiopia. *Ethiop.J.Health Sci.,* 19(1):1-7.

Shimizu, Y., Matsuto, S., Mizunama, Y. and Okada, I. (1970a). Studies on the flavours of roasted barley. Part V. Further separation and identification of carbonyl compounds. *Agric. Biol. Chem.,* 23, 276-279.

Shimizu, Y., Matsuto, S., Mizunama.Y. and Okada, I. (1970b). Studies on the flavours of roasted barley. Part VI. Separation and identification of flavour compounds. *Agric. Biol. Chem.,* 34(6), 845-851.

Shobana S., and Malleshi, N.G. (2007). Preparation and functional properties of decorticated finger millet *J.Food Eng.,* 79, 529-538.

Shobana, S., Sreerama, Y. N and Malleshi, N. G. (2009). Composition and enzyme inhibitory properties of finger millet (*Eleusine coracana)* seed coat phenolics: mode of inhibition of α-glucosidase and pancreatic amylase. *Food Chem.,* 115(4), 1268-1273.

Shobana, S., Harsha, M.R., Platel, K., Srinivasan, K. and Malleshi, N.G. (2010).Amelioration of hyperglycemia and its associated complications by finger millet (*Eleusine coracana).* Seed coat matter in streptozotocin induced diabetic rats. *Bri.J. Nut.,* 104(12), 1787-1795.

Shukla, S.S., Gupta, O.P., Sawarkar, N.S. and Sharma, Y.K. (1985). Study of Macro and micro mineral nutrient contents of some cultivars of Ragi (*Eleusine coracana Gaertn). The Ind. J. Nutri Diet.,*22: 249-252.

Shukla,S.S., Gupta,O.P., Sharma,Y.K. and Sawarkar, N.S. (1986). Puffing quality characteristics of some Ragi cultivars. *JFST,* 23, 329-330.

Shurpalekar, S.S., Joseph, A.A., Moorjani, M.M., Lahiry, N.L., Indiramma, K., Swaminathan, M., Sreenivasan, A. and Subrahmanyan, V. (1962a). Supplementary value of fish flour fortified with vitamins to poor Indian diets based on different cereals and millets. *Fd.Sci.Mysore.* 11, 49-51.

Shurpalekar,S.S., Joseph,A.A.,Lahiry,N.L.,Moorjani,M.M.,Sankaran,A.N., Swaminathan, M., Sreenivasan, A. and Subrahmanyan, V. (1962b). Supplementary value of fish flour and a food containing low fat groundnut flour, Bengalgram flour and fish flour to poor rice diet. *Fd.Sci.Mysore.*, 11, 45-48.

Shyama Prasad Rao, R; Sai Manohar, R; Muralikrishna, G (2007) Functional properties of water-soluble non-starch polysaccharides from rice and ragi: effect on dough characteristics and baking quality. *LWT.*, 40(10), 1678-1686

Singh, V and Ali,. S.Z (2000) Acid degradation of starch. The effect of acid and starch type. *Carbo. Polymers.*, 41(2), 191-195.

Siucela, M., Taylor, J.R.N., Williano, W.A.J.de Dnodu, K.G. (2007). Occurrence and location of tannins in finger millet grain and antioxidant activity of different grain type. *Cereal Chem.*, 84(2), 169-174.

Slade, L. and Levine, H. (1998). Non equilibrium melting of native granular starch Part-I. Temperature location of the glass transition associated with gelatinisation of A-Type cereal starches. *Carbohydrate Polymers.*, **8**, 183-208.

Slavin, J.L. (2005). Dietary fibre and body weight. *Nutrition.*, 21, 411-418.

Soana, J. K.D I, Hosana, A., Tomomatsu A., Kactoh, K and Matsuyama, A. (1982). Finger millet and its utilisation for manufacture of fermented foods in Indonesia. *J. Jap. Soc. Food. Sci. Technol.*, 29(11, 685-692.

Solpico, F.O. and Yambao, A.N. (1966). Performance test of millet at the economic garden *J.Pl.Ind.*, 31, 219-229.

Sonnad, S.K. (2005). *Stability analysis in white ragi* (*Eleusine Coracana Gaertn.*) genotypes. M.Sc. thesis submitted UAS, Dharwad.

Sperling, L. H. (1994). *Introduction to physical polymer science.* LH Sperling Ed., Wiley Interscience., New York, 1994, 224-295.

Sreedhara Murthy, S.; Avadhani, K. K.; Chand, N. and Srinivas, T (1997) Factors influencing grain morphology in finger millet. [*Eleusine coracana* (L.) Gaertn.]. *Tropical-Agriculture.*, 74(4), 254-259.

SreeRamulu, U.S. and Maria, Kulandai, A. (1964). The composition of the ragi (*Eleusine coracana)* grain and straw as affected by the application of farmyard manure and superphosphate fertilizer. *Madras Agric.* J., 51, 379-385.

Sridhar, R. & Laxminarayana, G. (1994). Lipid class contents and fatty acid composition of small millets: Little (Panicum sumatrense), Kodo (Paspalum scrobiculatum), and Barnyard (Echinocloa colona). *J. Agric. Chem.*, 40, 2131-2134.

Sripriya, G., Chandrashekharan, K., Murthy, V.S. and Chandra, T.S. (1996). ESR spectroscopic studies on free radical quenching action of finger millet *(Eleusine coracana). Food. Chem.*, 57(4), 537-540.

Sripriya, G., Antony, U. and Chandra, T.S. (1997). Changes in carbohydrates, free amino acids, organic acids, phytate and HCl extractability of minerals during germination and fermentation of finger millet (*Eleusine coracana*). *Food. Chem.*, 58(4), 345-350.

Stone, M and Wright, B.(1980) *Knox's Cube Test*. Chicago: Stoelting Company.

Subba Rao, M.V.S.S.T and Muralikrishna, G. (2001). Non starch polysaccharides and bound phenolic acids from native and malted finger millet (ragi *Eleusine coracana* INDAF 15). *Food Chem.*, 72, 187-192.

SubbaRao, M.V.S.S.T. and Muralikrishna, G (2002). Evaluation of the antioxidant properties of free and bound phenolic acids from native and malted finger millet (Ragi, *Eleusine coracana*, Indaf-15). *J. Agric. Food Chem.*, 50, 889-892.

SubbaRao, M.V.S.S.T., Sai Manohar, R. and Muralikrishna, G. (2004). Functional characteristics of non starch polysaccharides obtained from native (n) and malted (m) finger millet (ragi, *Eleusine coracana*, INDAF-15). *Food Chem.*, 88, 453-460.

Subrahmanyan,V., Murthy, H.B.N. and Swaminathan, M. (1954a). Effects of partial replacement of rice, wheat or ragi (*Eleusine coracana)* by tuber flours on the nutritive value of poor vegetarian diets. *Br.J.Nutr.*, 8, 1-10.

Subrahmanyan, V., Krishnamurthy, K., Swaminathan, M., Bhatia, D.S. and Raghunatha Rao, Y.K. (1954b). Supplementary value of cottonseed flour to wheat and ragi diets. Bull.cen.Fd.technol.Res.Inst., *Mysore.* 3, 225-226.

Subrahmanyan, V., Narayana Rao, M., Rama Rao, G. and Swaminathan, M. (1955). The metabolism of nitrogen, calcium and phosphorus in human adults on a poor vegetarian diet containing ragi. *Br.J.Nutr.*, 9, 350-357.

Sudha, M.L., Vetrimani, R. and Rahim, A. (1998). Quality of vermicelli from finger millet (*Eleusine coracana)* and its blend with different milled wheat fractions. *Food Res. Int.*, 31(2), 99-104.

Sultana, A. and Geervani, P. (1981). An assessment of the protein quality and vitamin B content of commonly used fermented products of legumes and millets. *Journal of the Science of Food and Agriculture.*, 32(8), 837–842.

Sundararaj, D.D. and Thulasidas, G. (1993). *Finger millet: Botany of Field Crops,* 99-104 2nd editions, Macmillan India Ltd. Publishers.

Sur, G., Reddy, S.K., Swaminathan, M. and Subrahmanyan, V. (1954a). Supplementary value of food yeast *(Torula utilis)* to poor vegetable diets based on cereals. *Bull.cen.Fd.technol.Res.Inst, Mysore.* 3, 111-112.

Sur, G., Reddy, S.K., Swaminathan, M. and Subrahmanyan, V. (1954b). Supplementary value of the proteins of food yeast cereal proteins. *Bull.cen.Fd.technol.Res.Inst, Mysore.* 4, 35-36.

Tasker, P.K., Rao, M.N. and Swaminathan, M. (1964). Supplementary value of a processed protein food based on a blend of coconut meal, groundnut flour and Bengalgram flour to poor Indian diets based on different cereals and millets. *J.Nutr.Diet.,* 1, 95-97.

Tatala, S., Ndossi, G., Ash, D. and Mamiro, P. (2007) Effect of germination of finger millet on nutritional value of foods and effect of food supplement on nutrition and anemia status in Tanzanian children. *Tanzan Health Res Bull.*, May. 9(2), 77-86.

Thapa S. and Tamang.J.P. (2004). Product characterization of kodo ko jaanr: fermented finger millet beverage of the Himalayas. *Food Microbio.,* 21(5), 617-622.

Tharanathan R.N. and Mahadevamma, S. (2003). Grain legumes – a boon to human nutrition. *Trends Food Sci. Tech.,* 14, 507-518.

Thatam, A.S., Fido, R.J., Moore, C.M., Kasarda, D.D., Kuzmicky, D.D., Keen, J.N. and Shewry, P.R. (1996). Characterisation of the major prolamines of Tef (Eragrostis tef) and Finger millet (*Eleusine coracana). J. Cereal Sci.,* 24, 65-71.

Theabaudin, J.Y., Lefebvre, A.C., Harrington, M. and Bourgeois, C.M. (1997). Dietary fibre: Nutritional and technological interest. *Trends Food Sci. Tech.,* 8, 41-48.

Thomas, M., Leelamma, S. and Kurup, P.A. (1990). Neutral detergent fibre from various foods and its hypocholesterolemic action in rats. *JFST.,* 27(5), 290-293.

Thompson, L.U. (1993) Potential health benefits and problems associated with antinutrients in foods. *Food Res. Int.,* 26, 131 -149.

Toeller, M. (1994). α-glucosidase inhibitors in diabetes: Efficacy in NIDDM subjects *Eur. J. Clin. Invest.,* 24, 31-35.

Tovey, F. I. (1994). Diet and duodenal ulcer. J. *Gastroenterol. Hepatol.,* 9, 177-185.

Tsugita, T., Kurate, T. and Fujimaki, M. (1978). Volatile components in the steam distillate of rice bran. *Agric. Biol. Chem.,* 42, 643-47.

Tyner, E.P., Lewis, H.B. and Eckstein, M.C. (1950). Niacin and the ability of cystine to augment deposition of liver fat. *J.biol.Chem.* 187, 651-654.

Udaya Kumar, M., Sashidhar, V.R and Prasad, T.G. (1989). Physiological approaches for improving productivity of finger millet under rainfed conditions. P 179-208. In: *Small Millets in Global Agriculture.* Ed: A

Seetharam, K.W. Riley and G. Harinarayana. Oxford & IBH Publishing Co. Pvt.Ltd., New Delhi.

Udayasekhara Rao, P. (1994). Evaluation of protein quality of brown and white ragi (*Eleusine coracana)* before and after malting. *Food Chem.,* 51, 433-436.

Udayasekhara Rao, P. and Deosthale, Y.G. (1988). In vitro availability of iron and zinc in white and coloured ragi (*Eleusine coracana):* Role of tannin and phytate. *Plant Food Hum. Nutr.,* 38, 35-41.

Usha, D and Malleshi, N. G. (2011). Changes in carbohydrates, proteins and lipids of finger millet after hydrothermal processing. *LWT Food Sci.* Technol., 44, 1636-1642.

Ushakumari, S.R., Rastogi, N.K., Malleshi, N.G. (2007). Optimisation of process variables for the preparation of expanded finger millet using response surface methodology. *J. Food. Eng.,* 82(1), 35-42.

Vasundhara, T.S. and Parihar, D.B. (1979). Studies on pyrazines of some roasted cereal flours. Zeitschrift Fur. *Lebensmittel-Unterschung and Forsung.,* 169(6), 468-71.

Venkannababu, B., Ramana, T. and Radhakrishnan, T.M. (1987). Chemical composition and protein content in hybrid varieties of finger millet. Ind. *J. of Agric Sciences.,* 57(7), 520-522.

Venkataraman, L.V., Becker, W.E., Khanum, P.M. and Murthy, I.A.S. (1977). Supplementary value of the proteins of alga *Scenedesmus acutus* to rice, ragi, wheat and peanut proteins. *Nutr.Rep.Int.,* 15, 145-155.

Venkataramana, R.S. and Krishna Rao, D.V. (1961). Chemical composition and nutrient uptake of ragi. *J.Ind.Soc.Soil Sci.,* 9, 245-252.

Vidyavati, H. G., Mushtari Begum, J., Vijayakumari, J., Sumangala, S. Gokavi. And Shamshad Begum. (2004). Utilization of Finger millet in the preparation of *papad. JFST.,* 41(4), 379-382.

Virupaksha T.K., Geeta, R. and Nagaraju, D. (1975). Seed proteins of finger millet and their amino acid composition. *J.Sci. Food Agric.,* J26: 1237-1246.

Vishwanath, S. and Seetharam, A. (1989). Diseases of small millets and their management in India. In: *Small millets in global agriculture,* 237-253. Ed: A Seetharam, K.W. Riley and G. Harinarayana (eds). Oxford and IBH Publishing Co., New Delhi.

Viswanath, V., Urooj, A., Malleshi, N.G. (2009). Evaluation of antioxidant and antimicrobial properties of finger millet *(Eluesine coracana)* polyphenols. *Food Chem.,* 114, 340-346.

Wadikar, D.D., Vasudish, C.R., Premavalli, K.S. and Bawa, A.S. (2006). Effect of variety and processing on anti nutrients in finger millet. *JFST.*, 43(4), 370-373.

Wadikar, D.D., Premavalli, K.S., Satyanarayana Swamy, Y.S. and Bawa, A.S. (2007). Lipid profile of finger millet *(Eleusine coracana)* varieties. *JFST.*, 44(1), 79-81.

Wang, P., Kato, H. and Fijimiki, M. (1970). Studies on flavor components of roasted barley. Part IV: volatile sulphur compounds. *Agri.Biol.Chem.*, 43, 2425-2432.

Wang, P., Kato, H. and Fujimaki, M. (1968). Studies on flavor components of roasted barley. Part I. Production of flavor substances. *JFST* (Tokyo)., 15(11), 514, 1968 From FSTA, 3, 1970.

Wang, P., Kato, H. and Fujimaki, M. (1969). Studies on flavor components of roasted barley. III Major volatile basic compounds. *Agric. Biol. Chem.*, 33, 1775-81.

Wang, P.S. and Odell, G.V. (1973). Formation of pyrazine from thermal treatment of some amino hydroxyl compounds. *J. Agric. Food Chem.*, 21, 868-71.

Wankhede (1979): Carbohydrate composition of finger millet and foxtailmillet. *Plants Foods for Human Nutrition.*, 24 (4), 293 -303.

Wealth of India.(1952). *Eleusine coracana* Vol.3,161-169.

Withycombe, D.A., Lindsay, R.C. and Stuiber, D.A (1976a). Isolation and identification of volatile components from wild rice grain. *J. Agric. Food Chem.*, 26, 816-818.

Woo and Seib, P. A. (2002). *Cross-linked resietant starch: Preparation and properties, cereal chem.*, 79(6), 819-825.

World Health Organization. *Research and Evaluation of Traditional Medicine. Ed. Ranjit Roy Chaudhury and Uton Muchtar Rafei,* Korea. WHO 2005.

Yajima, I.T., Yanai, M., Nakamura, H., Sakakibara and Hayashi, K. (1979). Volatile flavor components of cooked kaorimai (scented rice). *Agric. Biol.Chem.*, 43, 2425-29.

Yeol Baik, M. K., Joong Kim, K., Cheol Cheon, Y., Chutta and Kim. W. S (1997). Recrystalization kinetics and glass transition of rice starch gel system: *Journal of Agricultural Food Chemistry.*, 45, 4242-4248.

Yoo, S.S. and HO, C.T. (1997). Ryrazine generation from the Maillard reaction of mixed amino acids in model system. *Perfume. Flavor.*, 22, 49-52.

Yue, P. and Waring, H. (1998). Functionality of resistant starch in food applications. *Food Australia.*, 50(12):615-621.

Zake,V.M and Khizzah, B.W. (1989). *Finger millet improvement in Uganda* p 137-148. In: Small millets in global agriculture.

Zvauya, R., Mygochi, T & Parawira, W. (1997). Microbial and Biochemical changes occurring during production of masvusvu and mangisi, traditional Zimbabwean beverages. *Plant Food Hum. Nutr.*, 51(1), 43-51.

INDEX

D

G

H

I

N

O

P

Q

R

S

T

U

V

W

X

Y

Z